ZINC AND ITS ALLOYS AND COMPOUNDS

ELLIS HORWOOD SERIES IN INDUSTRIAL METALS

Series Editor: E. G. WEST, OBE, Metallurgical Consultant, and former Director of the Copper Development Association, London

TIN AND ITS ALLOYS AND COMPOUNDS
B. T. K. BARRY and C. J. THWAITES, International Tin Research Institute, Middlesex
COBALT AND ITS ALLOYS
W. BETTERIDGE, former Chief Scientist, International Nickel Limited, UK
NICKEL AND ITS ALLOYS
W. BETTERIDGE, former Chief Scientist, International Nickel Limited, UK
PRINCIPLES OF HYDROMETALLURGICAL EXTRACTION AND RECLAIMATION
E. JACKSON, Principal Lecturer in Metallurgy, Sheffield City Polytechnic, Sheffield
ZINC AND ITS ALLOYS AND COMPOUNDS
S. W. K. MORGAN, former Managing Director, Imperial Smelting Processes Ltd.
COPPER AND ITS ALLOYS
E. G. WEST, OBE, former Director, Copper Development Association, London
BASIC CORROSION AND OXIDATION
J. M. WEST, Department of Metallurgy, University of Sheffield

ZINC AND ITS ALLOYS AND COMPOUNDS

S. W. K. MORGAN, B.Sc., A.R.S.M.
former Managing Director
Imperial Smelting Processes Limited

ELLIS HORWOOD LIMITED
Publishers · Chichester

Halsted Press: a division of
JOHN WILEY & SONS
New York · Chichester · Brisbane · Toronto

First published in 1985 by

ELLIS HORWOOD LIMITED
Market Cross House, Cooper Street, Chichester, West Sussex, PO19 1EB, England

The publisher's colophon is reproduced from James Gillison's drawing of the ancient Market Cross, Chichester.

Distributors:

Australia, New Zealand, South-east Asia:
Jacaranda-Wiley Ltd., Jacaranda Press,
JOHN WILEY & SONS INC.,
G.P.O. Box 859, Brisbane, Queensland 4001, Australia

Canada:
JOHN WILEY & SONS CANADA LIMITED
22 Worcester Road, Rexdale, Ontario, Canada.

Europe, Africa:
JOHN WILEY & SONS LIMITED
Baffins Lane, Chichester, West Sussex, England.

North and South America and the rest of the world:
Halsted Press: a division of
JOHN WILEY & SONS
605 Third Avenue, New York, N.Y. 10158 U.S.A.

669.5 moR

©*1985 S.W.K. Morgan/Ellis Horwood Limited*

British Library Cataloguing in Publication Data
Morgan, S.W.K.
Zinc and its alloys and compounds. –
(Ellis Horwood series in metal science)
1. Zinc   2. Zinc alloys
I. Title
669'.9652   TN796

Library of Congress Card No. 85–5590

ISBN 0–85312–762–X (Ellis Horwood Limited)
ISBN 0–470–20213–0 (Halsted Press)

Typeset by Ellis Horwood Limited
Printed in Great Britain by Unwin Brothers of Woking

Table of Contents

Author's Preface

Whilst zinc ranks as one of the common metals, it has a number of properties which combine to differentiate it sharply from its fellows. It has a relatively low boiling point ($927°C$), and since its oxide is not reduced by carbon until temperatures above this value are applied, the metal is produced as a vapour, and the smelting methods which were developed early in metallurgical history for the production of copper, iron, lead, and tin could not be used for zinc. Its commercial production had to await the development of a retort method to vaporise the zinc, and condense it separately. As a consequence, metallic zinc was not produced in quantity until much later than the other common metals, but a number of solutions to the problem were eventually found, and at least six radically different methods of production have been developed, and have had wide commercial application.

As would be expected from its low melting point of $419°C$, unalloyed zinc has only mediocre physical properties, and its use in engineering applications is consequently restricted. However, it is used extensively in alloys: for example, with aluminium (4 per cent) it forms a series of excellent die-casting alloys, which are used in lightly stressed applications in many fields. Alloys with higher aluminium content (8 per cent) have been developed which have some application in gravity casting. The addition of up to 45 per cent of zinc to copper forms the important and long-established series of brass alloys — perhaps zinc's major contribution to engineering materials.

The resistance of zinc to prolonged stress is not high and the metal is prone to creep, but alloys with small copper and titanium additions have been developed which are much more creep-resistant. Certain alloys are able to deform very readily through the so-called phenomenon of superplasticity.

Perhaps the most important application of zinc arises from its chemical properties. Its resistance to atmospheric corrosion is high, and adherent coatings of zinc on iron can be produced which have many times the corrosion-resistance of the basic steel. As zinc is electronegative to iron it tends to corrode preferentially, further protecting the steel surface.

Zinc plays a useful, if prosaic, role in many fields. Although most applications of the metal are under pressure from possible substitutes, the demand continues to grow, and there can be little doubt that it will be used in quantity for many years to come. As with other basic raw materials the reserves available in the earth's crust are finite. As far as ore reserves are concerned zinc is better placed than the other common metals. If demand continues at the present rate, however, the availability of sufficient high grade concentrates, on which the industry now mainly depends, will be threatened. There is need for the development of extractive techniques able to treat the large quantities of low grade deposits, which cannot at present be used economically.

The book provides basic information on zinc extraction including economic factors, the properties of zinc, and the major commercial applications of the metal.

The author is grateful for the considerable amount of help received from friends in the zinc industry. He is particularly indebted to his ex-colleague, A. W. Richards, for much general help and criticism. Tony Chivers of the Zinc Development Association has been of great assistance in the section on the applications of the metal. The section on the electrolytic process owes much to Amel Gordon of the Electrolytic Company of Australasia, and that on the geology of zinc deposits to Dr Edwards of the Camborne School of Mines. The author was particularly fortunate to be able to call on A. K. Barbour's wide knowledge of environmental and hygiene problems and John Callaghan's refluxing experience. Information was also supplied by other Rio Tinto Zinc colleagues at Avonmouth — Colin Harris, Bill Hopkin, A. W. Robson and Hayden Monk. Assembly and typing was in the skilled hands of June Wheeler.

Thanks are due to Paul Craddock of the British Museum for a preview of the exciting information on early zinc smelting in India, which their recent expedition to Zawar has discovered.

The author is also indebted to Dr E. G. West — the general editor of this series — for his criticism and advice.

S.W.K.M.

1

History and general properties

The production of zinc occurred much later than that of the other common metals. Whilst copper was smelted from its ores probably about 5000 BC, lead produced about 4000 BC and iron about 2000 BC, zinc does not seem to have been available on a commercial scale until the fourteenth century AD.

As a metal, zinc was certainly known before this time. Brass, an alloy of zinc and copper, was produced by the Romans as early as 200 BC, but the method they used involved the heating together in crucibles of copper, zinc oxide and carbon. The zinc formed by reduction of the oxide was absorbed immediately in the copper, and was not produced as a separate phase.

Small quantities of metallic zinc were occasionally recovered from the flues of lead-smelting furnaces. Most lead ores contain zinc, and during smelting, a proportion of the zinc oxide present is reduced and forms zinc vapour, and whilst most of this is oxidised immediately in the upper levels of the furnace, occasionally some escapes and condenses as metal in the flues, and can be recovered from the fume scraped from the walls. Only very small amounts could have been collected in this way, insufficient to have any commercial impact, but enough to show the existence of the metal, to enable its characteristics to be determined, and possibly to account for the few zinc objects which archaeologists have found dating from before the Christian era.

1.1 THE EARLIEST PRODUCTION METHODS

When zinc metal was first made as a separate entity is not known. A much-discussed quotation from the Greek geographer Strabo (63 BC —AD 24) states 'At Andeira there is a stone which when burnt becomes iron. It is then put into a furnace, together with some kind of earth, when it distills a mock silver (*pseudogyrum*)'. This could be claimed to describe the production of zinc, but some philologists are uncertain that Strabo used the word 'distills' in its modern sense [1]. Whatever importance is attached to Strabo, it would seem likely that some of the early metallurgists did produce the metal, albeit on a laboratory scale. Mercury was produced by distillation and condensation before the Christian era, and whilst the production of zinc is more difficult, involving much higher temperatures, the technique required is similar.

Fig. 1 – Section through an ancient retort for zinc making, found at Zawar in India.

The very considerable problems involved in the commercial production of zinc were first overcome in the East. There is clear evidence that in India at Zawar in Rajasthan the metal was produced in quantity in the fourteenth century, and perhaps even earlier. At Zawar, near mining operations recovering lead and zinc, which are still being conducted by Hindustan Zinc Limited, there are dumps containing large numbers of small clay retorts, which were used for zinc distillation. The main bulk of the retorts are cigar-shaped, approximately 26 cm long and 15 cm in diameter. A section is shown in Fig. 1. It would appear that the body of the retort was first moulded by hand from clay made of local soils, and was then filled with a charge of a mixture of zinc oxide and charcoal. A clay cap was fitted to cover the retort, and this cap was pierced by a clay tube 1.8 cm in diameter to carry the vapours out of the retort.

For some time, little was known of the method used in Zawar, but in late 1983 an archaeological examination on the site was mounted by the British Museum, Hindustan Zinc and the University of Baroda, which discovered the remains of several furnaces. A photograph of one of these is shown in Fig. 2.

Fig. 2 – Remains of a zinc distillation furnace at Zawar.
(Courtesy of Dr Paul Craddock, British Museum.)

FIRE BOX

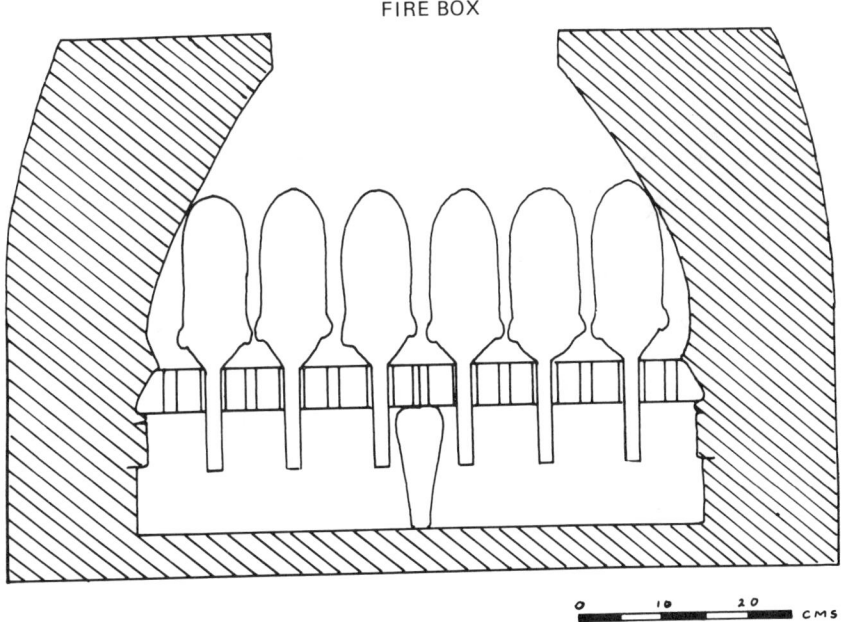

Fig. 3 – Arrangement of retorts at Zawar.

So far only preliminary descriptions of the existing investigation have been published [2], but the furnace described consisted of seven separate units in line. Each unit was in the form of a truncated pyramid approximately 70 cm square at the base, tapering to 40 cm at the top. The bottom of the main part of the furnace was formed by a perforated plate. The retorts were placed vertically downwards, with their necks protruding through holes in the plate. Figure 3 shows the arrangement diagrammatically.

It was not yet clear what system of condensation was used. It is possible that a separate collecting pot or crucible was fitted to the neck of each retort as it protruded through the bottom plate. This would provide adequate condensation conditions but it would have been extremely cumbersome, and few remains of the vast number of collecting vessels which must have been used have yet been found. An interesting possibility is that no separate condensing vessels were used but the whole space below the plate formed a common condenser. This principle was adopted in the horizontal furnaces at Overpelt in the 1950s as described on p. 63. An argument against the possibility that this method was used at Zawar is the large number of small holes in the bottom plate (other than those through which the necks of the retorts protruded). However, if the section below the bottom plate was sealed, when distillation commenced a positive pressure from the carbon monoxide formed in the retorts would soon build up. This would be forced up into the combustion zone surrounding the retorts in the upper zone, and entry of combustion gases into the lower condensing section (where they would be fatal) would be barred. The holes above the plate would be covered by fuel, so that again entry of combustion gases into the condensing chamber would be hindered.

Further investigation on the site is required to determine the type of condensation used. If however, as speculated above, the chamber below the retorts was used as a common condenser, then the Zawar metallurgists deserve credit for developing an important simplification to distillation practice, which was not rediscovered until the work at Overpelt four hundred years later.

The retorts were only used for one distillation cycle. Presumably, after one such cycle the furnace would be allowed to cool, the upper part dismantled, and the beads of zinc metal collected in the condensing section would be recovered and melted down. The retorts containing their spent charge would be removed and the floor plate cleaned and repaired. In the meantime a new batch of retorts would have been filled with charge, the clay cap added and sealed. They would be placed in position, the collecting vessels fitted (if used) and the next cycle would begin.

Each distillation period must have taken at least 8 days and perhaps longer [5]. This would have been made up as follows:

Cooling and dismantling	2 days
Charging, drying and heating to distillation temperature	4 days
Distillation	2 days

The heating of the cold retorts must have been gradual to avoid cracking of the refractories. A lengthy distillation period is essential. To reduce zinc oxide, a good deal of heat is required — much more than is the case with lead or copper — and this heat must be forced through the retort wall — a poor conductor. It must then penetrate into the centre of the charge to complete reduction, and this is time consuming. Reduction of zinc oxide by carbon does not begin until temperatures of $1000°C$ are reached, so that the inside of the retort must be raised to, and maintained at, this temperature. At Zawar they must have used charcoal as fuel, and to reach temperatures as high as $1200°C$ throughout the whole furnace laboratory, using cold air, would be a creditable performance, and about the maximum possible. Even with such temperatures the transfer of heat through the walls of the retort (which governs the distillation rate) would be slow, since the thermal gradient available would be only $200°C$. The horizontal furnaces at Avonmouth, using relatively thin-walled retorts, made of high grade Belgian refractory clays, had a heating period of 18 hours, but used furnace finishing temperatures of $1360°C$. In the USA, with retorts containing a high proportion of carborundum (to improve heat conductivity) a 48-hour cycle, with finishing temperatures of $1320°C$, was used. In the English process based on Champion's development at Warmley, and using a much more sophisticated furnace than those used at Zawar, a distillation period of 3 days was employed.

The internal volume of a Zawar retort was of the order of 1 litre. By analogy with the horizontal retort practice at Avonmouth (and assuming the same recovery of metal of 75 per cent) each Zawar retort would have produced 0.6 kg of zinc metal per cycle. The annual capacity of a 7-unit furnace as described by Craddock, assuming 48 cycles per year, would then be of the order of 7000 kg. The blast furnace at Avonmouth produces this amount in approximately 20 minutes.

The requirement for charcoal must have been considerable. The English process, based on Champion's development, using a fairly sophisticated furnace developed for the glass industry, required 22 tons of carbon to produce 1 ton of zinc metal. This was due to the necessity to hold the furnace at high temperature for a long time, to force enough heat through the retort wall to effect reduction. It is unlikely that at Zawar the proportion of fuel used was less. Assuming the 1:22 ratio, the 7-unit furnace producing 150 kg of zinc would require 3300 kg of charcoal per cycle.

Craddock claims that radiocarbon dating places the operation in the sixteenth century [2]. As far as is known this is the first example of commercial production of zinc, and the early metallurgists at Zawar deserve great credit.

In China, as in India, the production of zinc was practised before the necessary technique was known in the West. The production of brass in China was known in the third century BC, but the alloy was almost certainly made by the cementation process, also used by the Romans, in which zinc is not produced as a separate phase. From philological evidence quoted by Dr Needham [3] it is possible that the Chinese were aware of and could possibly produce the metal before the Christian era, but there is no other evidence to confirm this, and it seems unlikely that commercial zinc production in China predated the Zawar operations.

Unfortunately, as yet nothing is known of the method used first in China. In 1925 Wheler described an early Chinese method of producing zinc [4]. He states clay retorts were used 31 cm long and 12 cm diameter, and these were filled with a loosely packed mixture of roasted ore and anthracite. To the top of each retort was first luted a clay cup, to act as a container for the zinc subsequently condensed, and then a crucible-shaped extension which was covered with an iron disc. A number of such retorts were placed in a furnace, packed round with fuel and heated. When the temperature inside the retort reached 1000°C, reduction of the zinc oxide commenced, and zinc vapour and carbon monoxide passed up the retort into the cooler chamber above, where most of the zinc condensed, and collected in the clay cup. A reconstruction of the retort and condensation system described by Wheler is shown in Fig. 4.

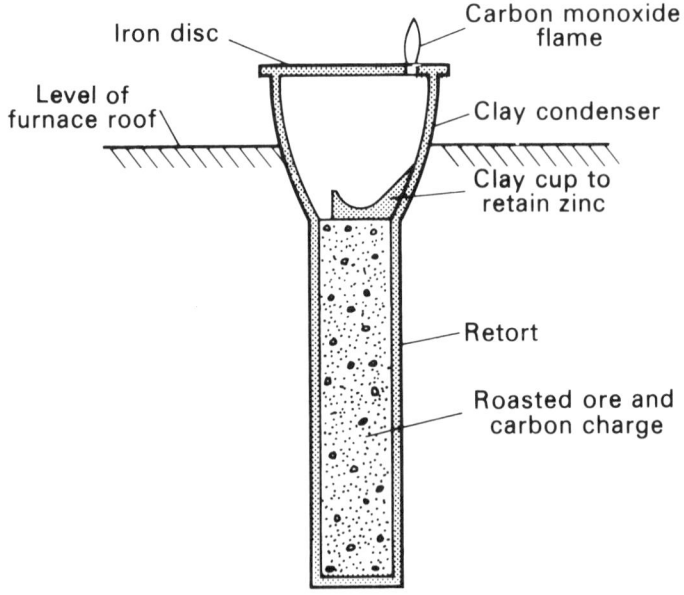

Fig. 4 – Diagrammatic reconstruction of an early Chinese retort and condensing system.

Unfortunately Wheler does not state when the method was first operated. The furnace he describes is relatively sophisticated and must have been the product of considerable development. Some further details of the same Chinese method are given by R. F. Tylecote [5].

The fact that in this method the vapour rose upwards through the retort to condense at the top is of particular interest, since in the method used at Zawar (and in that later developed by Champion in Bristol) distillation occurred downwards. It is tempting to assume that the early Chinese metallurgists used the principle of upward distillation which Wheler describes, but this is supposition. If it were so, however, the first Chinese and Indian solutions to the difficult problem of large-scale zinc production must have been developed independently.

1.2 THE DEVELOPMENT OF THE ZINC INDUSTRY

The first commercial plant to produce metallic zinc in Europe was built and worked at Warmley, near Bristol, by William Chapmion, who by 1747 was originator and master of a large operation. He based his new process on the principle of downward distillation, as used at Zawar. Its development was a major breakthrough in non-ferrous metal extraction, and the metallurgists who first used it to produce zinc commercially deserve praise. There is no evidence that Champion knew of the Zawar operation. When campaigning for an extension of his patent in 1750 he claimed he travelled much in Europe before he developed his process, but he makes no mention of travelling further East. There must have been considerable speculation amongst metallurgists in the West as to how the metal was produced in India and China, and it seems unlikely that some description of the downward distillation principle had not reached Europe. What is certain is that Champion took the principle (whether he rediscovered it or not), and built round it a technically successful large-scale operation, far in advance of the cumbersome small-scale furnaces used at Zawar.

One of Champion's achievements was that he saw the potential of, and adopted, a furnace which was already in use for making glass. By the end of the seventeenth century a flourishing glass industry had already been established in Bristol, with a number of these furnaces in operation. Each was circular, with a conical roof, and the glass was melted in large clay crucibles about 1 m high and 0.9 m in diameter. Each furnace generally held six such crucibles placed on a circular base and heated directly by a coal fire from below. With skilful stoking, and utilising the good draught which the tall furnace provided, the crucibles could be heated to temperatures in excess of $1000°C$.

Champion saw that the crucibles would provide the retorts that he needed, and that the furnace would enable him to reach and maintain the temperature required for distillation.

A further step forward Champion made in the furnace he built, was the provision of efficient condensation. He drilled a hole in the crucible bottom and fitted in it an 18 cm diameter wrought iron pipe which passed down through the floor of the furnace into the cooler cellar or 'cave' below. A charge of zinc oxide (formed by calcining calamine, the zinc ore found on the Mendips) and lump charcoal was placed in the crucible, and the temperature of the furnace was

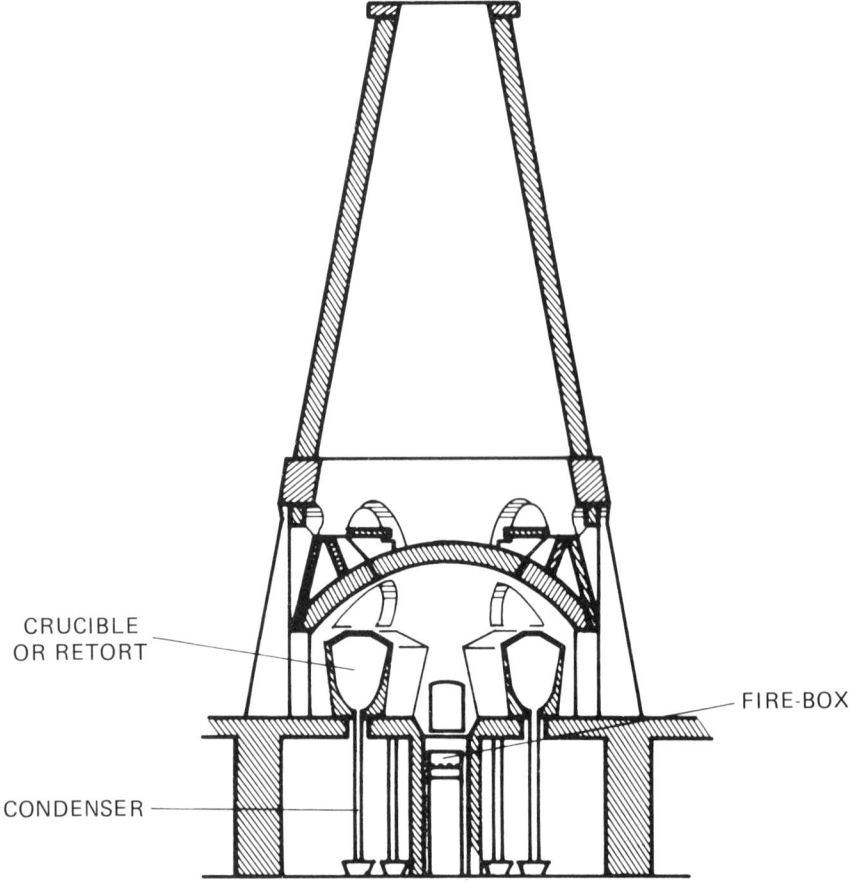

CRUCIBLE
OR RETORT

FIRE-BOX

CONDENSER

Fig. 5 – Diagram of English furnace for zinc production, based on Champion's
development at Warmley.

raised to a white heat. Reduction of the zinc oxide began, and the zinc vapour
formed was forced, by the pressure built up in the crucible, down through the
hole in the bottom into the iron tube, where the temperature dropped and
condensation occurred.

The next advance was made in Silesia. In 1798, some fifty years after
Champion's first success, Johann Ruberg in Wessola produced quantities of
zinc, again in a glass-making furnace. In this furnace he used fireclay muffles
— 1000 mm long, 500 mm wide and 600 mm high — and he fitted four such
muffles inside the furnace. The open end of each muffle just protruded through
the furnace wall and was closed by means of a detachable clay slab containing a
hole, as shown diagrammatically in Fig. 6.

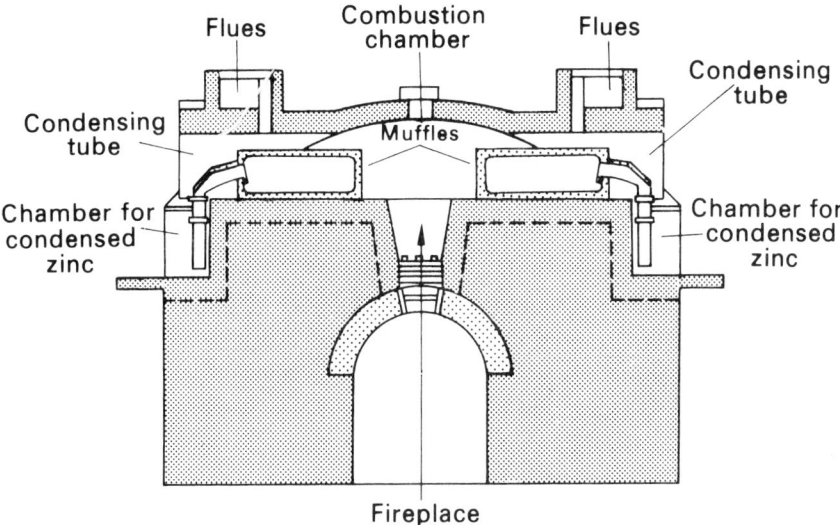

Fig. 6 — Diagram of Silesian furnace for zinc smelting, as used by Ruberg.

As the source of zinc, Ruberg first used zinc oxide collected in the flues of iron blast furnaces. This material was then reasonably pure and plentiful, since the local iron ores contained zinc, which was volatilised during the iron smelting, but, later, calcined calamine was used. The furnace charge of zinc oxide mixed with lump charcoal was packed into the muffles and the clay and slab replaced. The condensers were fitted and the temperature raised, to produce zinc.

Although Ruberg gained little personal benefit from his work — he died in 1808, apparently in poverty and an alcoholic — the method he pioneered was considerably extended. Furnaces contained as many as 24 muffles of increased size, and in the nineteenth century, using Ruberg's method, Silesia became one of the world's largest centres of zinc production. The process, however, gradually gave way to the Belgian method developed in Liège by the Abbé Dony in 1808, using a greater number of smaller retorts which were easier to make and handle.

Jean Jaques Daniel Dony was born in Liège in 1759. He was known as the Abbé Dony, although it is not clear that he actually took orders; but he studied both theology and chemistry. Little is known of his theological work, but he made a close study of the problem of producing metallic zinc from the local calamine, and in 1808, the year Ruberg died, he successfully produced zinc in quantity. It was of good quality since it was malleable and could be rolled into sheet.

Dony's great contribution to zinc smelting lay in the fact that he developed a method using horizontal retorts, superior to that pioneered by Ruberg. His achievement was due to the fact that, in the Liège area, some of the best re-fractory clays in the world were available. This enabled him to construct long retorts which could be placed horizontally in a furnace supported only at each end, but which were so strong that they could be heated to, and maintained at, the high temperatures necessary for reduction, without sagging. They could be placed in rows in a furnace and heated from all sides. He used retorts about 1000 mm long and 150 mm in diameter, which were smaller than Ruberg's and could be fashioned and handled more easily. The open end, or mouth of each retort, protruded through the front wall of the furnace, and the retorts could be charged and discharged without difficulty, and the retort used many times. After charging, a fireclay, fish-tail shaped clay condenser was fitted to each retort, which cooled and condensed the zinc from the vapours leaving the retort. It could be removed at the end of each cycle, the spent residues removed, and the retort refilled with fresh charge.

The method, though basically similar to that of Ruberg, was an improve-ment. By using a large number of smaller retorts set in rows in a furnace, Dony could provide more heating area per unit of charge, and required less fuel per unit of zinc produced.

Like Ruberg, Dony himself benefited little from his achievement. His process was taken over by others more commercially minded and he died in 1819 in 'profound misery'. Similarly, Champion gained little materially from his pioneering work at Warnley as he was declared bankrupt in 1767, and until he died in 1795 is said to have worked as a mason.

In 1916 the full scale development of the electrolytic process began, and this was a radical departure from the thermal methods, which had previously held the field. In 1929 the New Jersey Zinc Company perfected the vertical retort process – the first continuous system of zinc distillation. In 1935 the St. Joseph Lead Company successfully modified the vertical retort method by heating the charge electrically, using its resistance to generate the necessary heat, and condensing the zinc vapour produced in a bath of molten zinc. In 1959 the first full scale zinc–lead blast furnace, developed at Avonmouth by the Imperial Smelting Corporation, came into operation at the Swansea Vale works of the company, using the lead splash condenser, which for the first time solved the problem of condensing zinc from a blast furnace gas. These processes will be described in more detail in Chapters 4, 5 and 6.

The history of zinc production is therefore complex, and a number of radically different solutions have been found to the basic problem of zinc oxide reduction. As late as 1975 five different processes were in use throughout the world, and the total world annual zinc production of the order of 5.5 million tonnes was made by the five processes in approximately the following proportions:

horizontal retort   5%
vertical retort   9%

electrothermal   7%
electrolytic   68%
blast furnace   12%

The pattern is changing rapidly however. The growing world energy shortage, and the necessity to avoid environmental pollution impose a series of new and pressing factors.

The horizontal process, with its high labour cost, poor efficiency, and the atmospheric pollution it presents is almost certainly obsolescent. It is doubtful whether any new vertical retort or electrothermic units will be built, since they require relatively high grade concentrates, and have high capital costs. In the immediate future, the choice for new installations would appear to be between the electrolytic route and the blast furnace, with the former preponderating. Many factors affect the choice, one of the major ones being the relative availability and cost of electric power and coke. However, if high grade concentrates become in short supply and lower grade materials are available, then the blast furnace will be favoured. This question is dealt with more fully in Chapters 5 and 6.

1.3 GENERAL PROPERTIES OF ZINC

Zinc is a silvery white metal with a relatively low melting point (419.5°C) and boiling point (907°C). When unalloyed, its strength and hardness is greater than that of tin or lead, but appreciably less than that of aluminium or copper. It cannot be used in stressed applications due to low creep-resistance. Except when very pure, zinc is brittle at ordinary temperatures, but malleable above 100°C, and can then be readily rolled. The metal is amenable to most forming operations. When alloyed with 4 per cent aluminium its strength and hardness is increased considerably, and such alloys have excellent castability and are widely used in die casting. Small additions of copper and titanium appreciably improve the creep-resistance of rolled sheet, producing a material of growing industrial importance. With additions of 20–22 per cent aluminium, superplasticity can be developed, the alloys flowing readily at temperatures of 220°C, under vacuum or other forming techniques. They can be readily moulded, yet regain normal strength and hardness at room temperature. A major application of zinc is as an alloying addition to copper, forming the range of brasses. By varying the proportion of zinc, alloys can be produced with a range of physical properties to meet a wide range of requirements.

One of the most useful characteristics of zinc is its resistance to atmospheric corrosion, one of its main applications thus being for the protection of steelwork. It holds a high position in the electromotive series of metals, below magnesium and aluminium, but above cadmium, iron, nickel and hydrogen, and thus a coating of zinc on steel will corrode preferentially in most media, and if discontinuities in the coating occur, sacrificial protection is provided. The electronegative character of zinc also leads to its use in considerable quantities in dry batteries.

1.4 CHEMICAL PROPERTIES OF ZINC

Zinc, cadmium, and mercury constitute Group IIB of the periodic table, and their electronic structure is shown in Table 1.

Table 1

The electron structure of the zinc group elements

	Atomic number	K	L	M	N	O	P
Zinc	30	2	8	18	2		
Cadmium	48	2	8	18	18	2	
Mercury	80	2	8	18	32	18	2

There are no natural radioactive isotopes of zinc. The normal valence states are $Zn(0)$ and $Zn(II)$. Compounds of $Zn(I)$ do not exist naturally, although ZnH and ZnX ($X = Br, Cl$) are known as spectrographic species. Zinc is generally divalent, and can give up two outer electrons to form an electrovalent compound, for example zinc carbonate $ZnCO_3$. It may also share those electrons as in $ZnCl_2$ in which the bonds are partly ionic and partly covalent.

Dry air has little attack on zinc at room temperatures, but above $200°C$ oxidation occurs rapidly. In moist air in the presence of carbon dioxide a hydrated basic carbonate is formed which adheres tenaciously, and is resistant to further action.

The metal is readily dissolved by most mineral acids, but zinc of high purity resists attack by dilute sulphuric acid due to hydrogen overvoltage — the electrolytic process is entirely dependent upon this fact. All grades of zinc are readily dissolved in strong caustic alkalis, both aqueous and fused, forming zincates, for example Na_2ZnO_2. In the form of dust or granules, zinc is widely used as a precipitant, and is a reducing agent for many ions such as ferric, manganate and chromate.

Zinc can form organic compounds such as ZnR_2 ($R = Me, Et, n$-Pr, n-Bu, n-pentyl, vinyl) etc. or $MeZnX$ ($X = Br, Cl$). Zinc also forms complexes with anionic, cationic or neutral ligands.

1.5 PHYSICAL PROPERTIES OF ZINC

The physical properties of zinc are given in Table 2, and the mechanical properties of the metal and its alloys are referred to in later chapters.

Table 2

Physical properties of zinc

linear coefficients of thermal expansion	polycrystalline	39.7×10^{-6} K^{-1} (293–523 K)
	a axis	14.3×10^{-6} K^{-1} (293–373 K)
	c axis	60.8×10^{-6} K^{-1} (293–373 K)
volume coefficient of thermal expansion		0.9×10^{-6} K^{-1} (293–673 K)
surface tension – liquid at melting point (692.7 K)	782 mN m^{-1}	
viscosity – liquid at melting point (692.7 K)	3.85 mNm^{-1}	
electrical resistivity	solid 293 K	5.96 $\mu\Omega$cm
	liquid 692.7 K	37.4 $\mu\Omega$cm
magnetic susceptibility per kg (diamagnetic, 293 K)	polycrystalline	-1.74×10^{-9}
	a axis	-1.55×10^{-9}
	c axis	-2.18×10^{-9}
atomic number	30	
atomic weight	65.38	
valence	2	

isotope abundance	mass number	64	66	67	68	70
	terrestrial %	50.9	27.3	3.9	17.4	0.5

crystal structure	hexagonal close packed	$a = 0.2664$ nm $c = 0.49469$ nm
		$c/a = 1.856$
twinning plane	1012	
atomic sizes	metallic radius	0.1332 nm
	ionic radius M^{2+}	0.075 nm
melting point	419.5°C (692.7 K)	
boiling point (101.335 kPa, 1 atm)	907°C (1180 K)	
density	solid, 25°C	7.14 g cm^{-3} (m kg m^{-3})
	solid, 419.5°C	6.83 g cm^{-3}
	liquid, 419.5°C	6.62 g cm^{-3}
	liquid, 800°C	6.25 g cm^{-3}
heat of fusion	7.28 kJ mol^{-1} at (692.7 K)	
heat of vaporisation	114.7 kJ mol^{-1} at (1180 K)	
heat capacity	solid Cp = 22.40 + 10.05 × 10^{-3} T J mol^{-1} (298–692.7 K)	
	liquid Cp = 31.40 J mol^{-1}	
	gas Cp = 20.80 J mol^{-1}	
thermal conductivity	solid (291 K)	113 W m^{-1} K^{-1}
	solid (692.7 K)	96 W m^{-1} K^{-1}
	liquid (692.7 K)	61 W m^{-1} K^{-1}
	liquid (1023 K)	57 W m^{-1} K^{-1}

modulus of elasticity	with zinc no strict proportionality between stress and strain exists, and under prolonged stress the metal is subject to creep. For short term loadings a value of 7×10^4 MN m^{-2} can be taken.

REFERENCES AND FURTHER READING

[1] Dawkins, J. M., *Zinc and Spelter. Notes on the Early History of Zinc*, Zinc Development Association, 1956.

[2] Craddock, P. T., Freestone, I. C., Gurjar, L. K., Hegde, K. T. M. and Sonawane, V. H., Early zinc production in India, *Mining Magazine*, Jan., p. 44, 1985.

[3] Needham, J., *Science and Civilisation in China*, Vol. 5, Part 2, p. 214, Cambridge University Press, London, 1974.

[4] Wheler, A. S., *Transactions of the Institution of Mining and Metallurgy*, Vol. 32, p. 266, 1923.

[5] Tylecote, R. F., Ancient metallurgy in China, *Metallurgist*, vol. 15, No. 9, p. 435, 1983.

[6] Smithells, C. J., *Metals Reference Book*, Butterworth, London, 1967.

[7] *The Properties of Zinc*, Consolidated Mining and Smelting Company of Canada, Trail, B.C., 1956.

[8] Kelley, K. K., High temperature heat content, heat capacity, and entropy data for inorganic compounds, Bulletin 476, US Bureau of Mines, 1949.

[9] Kelley, K. K., The free energies of vaporisation and vapour pressures of inorganic substances, Bulletin 383, US Bureau of Mines, 1935.

[10] Coughlin, J. P., Heats and free energies of formation of inorganic oxides, Bulletin 542, US Bureau of Mines, 1954.

2
Occurrence, concentration and economic factors

In total, vast quantities of zinc occur in the earth's crust. Fleming has pointed out [1] that taking the generally accepted value of 132 ppm for the zinc content of the earth's crust, the outer kilometre of the land mass alone contains 52×10^6 million tonnes of zinc. To the mining engineer this means little, since it is only when considerable concentration of the metal into workable ore deposits has taken place that profitable mining can proceed. At present prices and techniques, deposits containing less than 3 per cent zinc will rarely justify extraction, unless, as is often the case, metals such as lead, copper or silver are also present in the ore body and can be recovered simultaneously.

Fortunately, concentration into workable ore bodies has occurred on a fairly extensive scale in a number of areas of the earth's surface. Such zinc deposits tend to occur, not in isolation, but grouped together with some common association, generally explicable in terms of plate tectonics. The deposits almost always contain lead and cadmium, and minerals containing copper, silver and manganese are frequently present. Such associated groups or provinces are widely distributed over the earth's surface. Over 40 per cent of the world's zinc originates in Canada, the United States, Mexico and Australia, but important mines are found in Peru, Spain, Italy, Yugoslavia, Poland, Zaire, Japan, North Korea, Russia and Ireland.

2.1 THE GEOLOGY OF ZINC DEPOSITS

The geology of most zinc deposits is complex, and their exact mode of formation is often obscure, but in almost all cases, final concentration is thought to have occurred through hydrothermal mechanisms. Owing to crust disturbances occurring through movement at plate boundaries, aqueous solutions can be forced, sometimes at high pressures and temperatures, through porous strata, and in their passage dissolve zinc and other values present. On entering the marine environment the zinc, in the form of a soluble chloride complex, can be carried some distance along the sea bed, to be precipitated finally as zinc sulphide, when a source of hydrogen sulphide (sometimes produced bacterially) is encountered. The precipitates mix, and form beds contemporaneously with the sediments (usually of limestone or dolomite) already forming in the basin, and since they were laid down at the same time as the surrounding rocks, are known as syngenetic.

In certain cases the hydrothermal solutions with their load of zinc and other values may be passed through beds already lithified, and deposition will take place when conditions favourable to precipitation are met. Such deposits, since they were laid down after their host rocks, are known as epigenetic.

Since most mineral deposits and their host rocks have almost always been subjected to considerable metamorphosis after they have been laid down, it is not always possible to determine the chronology of deposition. As examples of syngenetic deposition the important Australian fields at Broken Hill and Mount Isa can probably be taken, since it would appear that mineralisation occurred during, or shortly after sedimentation. In the case of the Mississipi Valley type ores, it would seem that the zinc entered the dolomitised limestone some time after lithification, and the deposits are there epigenetic.

Zinc minerals
The common zinc containing minerals are listed in Table 3. Blende, marmatite and, to a lesser extent, calamine are the only minerals of any commercial significance today.

Table 3

Zinc minerals

Mineral	Formula
Zinc blende or sphalerite	ZnS
Marmatite	$(ZnFe)S$
Calamine or smithsonite	$ZnCO_3$
Hemimorphite	$4ZnO.2SiO_2.2H_2O$
Hydrozincite	$5ZnO.2CO_2.3H_2O$
Zincite	ZnO
Willemite	$2ZnO.SiO_2$

Associated minerals
Zinc blende almost invariably contains some cadmium, and practically all the cadmium used in commerce is produced as an important ancillary operation in zinc smelting. As almost all zinc ores also contain considerable quantities of galena (lead sulphide) most mines are producers of both zinc and lead. Chalcopyrite ($CuS.FeS$) is frequently present and the recovery of copper then adds to

the value of the ore, as does silver, which often occurs in small but profitable quantities. Gold is rarely present however. Calcite ($CaCO_3$) or dolomite ($CaMgCO_3$) and sometimes quartz (SiO_2) are the most common gangue constituents associated with blende. Fluorspar (CaF_2) and barite ($BaSO_4$) frequently occur, as do the manganese containing minerals, rhodocrosite ($MnCO_3$) and rhodonite ($MnO.SiO_2$).

2.2 ORE CONCENTRATION

The character of zinc ores is almost invariably highly complex, and in addition to containing considerable quantities of gangue material, almost always they contain other metals which must be recovered in as pure a form as possible. It is particularly important that high grade concentrates should be obtained, since in the case of both the electrolytic and blast furnace processes, the costs of production rise steeply with decreasing grade of concentrates treated, and therefore the maximum possible degree of separation and purification in the concentrating plant is always demanded. It is fortunate that the flotation process has been developed to a high pitch of perfection and good grade concentrates can now be obtained from most sulphide ore bodies. As a result flotation has now almost entirely replaced the older specific gravity based methods of concentration.

No two ore bodies are identical; they differ in mineralogical constitution and chemical composition. Even from the same mine considerable variations in the composition of the ore may occur from day to day. Except in the case of some mines, mainly in Eastern Canada, which have a low lead content, all zinc ore bodies contain sulphides of both lead and copper. The task facing the mill superintendent is to take this heterogeneous material which the mine gives him, and separate and collect at high efficiency the valuable minerals present, in their most saleable form, and in as high a degree of purity as possible. He is able to do this through the use of an often sophisticated programme, using various milling and flotation stages. In the latter he takes advantage of the fact that a variety of reagents have been produced which permit him to change at will the surface properties of the minerals he is separating.

Crushing and grinding

Methods of concentration must take account of the fact that the heterogeneous ore body consists of an aggregate of different minerals each with its own physical and chemical properties. The first stage is to crush and grind the ore to the extent that most of the discrete mineral particles are liberated from the matrix, and exist in the ground product in the free state. Only then can an efficient separation be made.

Minerals vary greatly in grain size, some being relatively coarse, for which crushing to 1 mm is sufficient to liberate them in a fairly pure state. There are exceptions, however, and most ores must be ground much finer to obtain satisfactory separation. With some, the finest degree of comminution commercially possible (less than 50 μm) is required to free the mineral particles.

Grinding is an expensive operation and the finer the particles required, the greater is the cost in power and the wear on the equipment. In addition, fine grinding increases the amount of slimes produced and, as this causes separation difficulties, every precaution is taken to avoid reducing the size of the crushed product beyond that necessary to liberate the valuable minerals. Thus most plants employ several grinding stages and use classifiers in closed circuit with the grinding mills (which are of either the rod or ball type).

Flotation of zinc sulphides
As stated in the previous section, the first stage in flotation (or in fact in any concentration process) is to crush the ore so that the matrix is broken up to such a degree that the sought-for minerals become individual grains.

Most mineral surfaces when freshly ground are wetted by water to a greater or lesser degree. The flotation stage rests basically on the principle that by the use of certain reagents the surface properties of the sulphide particles can be modified so that they become water-repellent and therefore tend to adhere preferentially to the air bubbles generated in the flotation cell. The sulphide particles thus collect in the froth and can be removed, and a separation from the gangue and other minerals can be made. Compounds which increase water repellency are known as collectors.

Alternatively, it is possible by the addition of other compounds to ensure that a mineral remains depressed in the cell. The reagents which prevent flotation are known as depressants.

Flotation practice varies from plant to plant due to the widely differing characteristics of the ore treated, but in general the practice followed is first to float copper sulphides, keeping the other minerals depressed, then lead, then zinc and sometimes finally pyrite, if this is economically worthwhile.

The reagents most frequently used to depress the zinc sulphide in such circuits are zinc sulphate and sodium cyanide. These permit copper and lead sulphides to be floated first, either separately or together. For the next stage, when it is desired to float the zinc sulphide, copper sulphate is added to activate the particles. This allows them to react with the collector so that, on aeration, a clean zinc sulphide concentrate can be made.

Flotation practice
Whilst the above represent broadly the principles employed in zinc flotation, in practice the circuits used are of considerable complexity. In order to obtain maximum efficiency of separation several stages of grinding and classification are used, interspersed with flotation stages with retreatment of both concentrate and tailings. Great care is taken to control the density and often the temperature

of the pulp and to ensure that the reagents used are added to greatest effect. Instrumentation and automatic control are extensively employed, as is continuous chemical analysis. The skills developed since the introduction of the flotation process at the beginning of the century have had a profound effect on the metallurgy of zinc, and have greatly increased the variety of ores which can now be successfully mined.

Flotation of zinc carbonate
As an instance of the versatility of the flotation process it has been found possible to float, and thus concentrate, zinc carbonate by the addition of sodium sulphide which presumably sulphidises to some extent the zinc carbonate surface. This practice, using collectors of the fatty amine type, was first developed in Sardinia.

2.3 THE STRUCTURE OF THE ZINC INDUSTRY

Since the distribution of ores is widespread, mining is carried out in a large number of countries. Owing to the efficiency of the flotation process, a high degree of concentration at the mine can be carried out, and concentrates containing 50–59 per cent zinc can be produced from most ores currently mined. These concentrates can be readily transported for treatment to centres which are rarely near the mines, but at locations which are determined by a number of factors, the most important being proximity to markets, availability of cheap power or fuel, and facilities for the disposal of sulphuric acid. The latter is of particular importance since with both the blast furnace and the electrolyte processes, the production of each tonne of zinc involves the production of two tonnes of sulphuric acid, which must be utilised.

Thus, in the past, smelting and mining have tended to develop as separate industries, although there are a number of large companies which operate both mines and smelters. A proportion of the latter are still of the custom type, treating a number of concentrates of different origins, bought on the open market. Again, although a few smelting companies have interests in die casting or galvanising operations to ensure outlets for their metal, most of the zinc-consuming industries, which establish the demand, are separate. The structure of the zinc industry as a whole is therefore complicated, without any high degree of integration.

Japan is an exception to this pattern. The smelters — both electrolytic and blast furnace — with an annual production capacity of some 800,000 tonnes are operated by six companies, which also control the internal mining capacity. These six companies are also closely associated with the trading corporations responsible for purchasing and marketing all the mineral products both at home and abroad. The association of the zinc industry with government agencies and the banking section is also close. The whole industry is therefore more tightly knit than in most other countries. The weakness of the Japanese position is the fact that, as is shown in Table 4, domestic mines can supply only one-third of the country's needs, and the remainder must be imported from abroad.

As is the case with other metals, competition for markets in the zinc industry is intense, and world wide. The economies of large-scale production are essential for survival and both the mining and smelting sections of the industry are composed almost entirely of large units. Thus the annual capacity of the electrolytic plant of Cominco at Trail, British Columbia, is claimed to be 300,000 short tons of zinc. The single blast furnace at Avonmouth of Commonwealth Smelting Limited is capable of producing 100,000 tonnes of zinc and 40,000 tonnes of lead per year. The industry is thus made up of a number of large companies, most with interests often extended over several continents.

There is a growing tendency in a number of countries to nationalise mining operations, and in certain cases smelter production. In both Mexico and Peru corporate control by foreign companies is giving way to public control of both mine and metal production. In India all production of zinc ore is in the public section, and the electrolytic plant at Udaipur is government-owned.

Table 4

1982 world zinc consumption, metal production and mine production by countries ('000 tonnes as metal)

	Consumption	Slab production	Mine production	Mine production less consumption
Europe	1522	1745	1007	−515
Japan	703	662	251	−452
USA	800	302	330	−470
Canada	120	512	1183	+1063
S. America, incl. Mexico	288	417	991	+703
Africa	177	214	303	+126
Australasia	112	295	662	+550
Asia and DPR Korea, less Japan and China	464	167	174	−290
Centrally planned economies, including China	1770	1643	1604	−166
Total	5956	5957	6505	

Source: World Bureau of Metal Statistics.

A factor with growing political and social implications is the geographical imbalance between zinc demand and ore supply, which is summarised in Table 4. This shows how heavily Europe, Japan and the United States already rely on imported concentrates or metal.

   In the case of the United States the position is likely to deteriorate still further unless a number of large new deposits can be discovered. In a report published by the US National Commission of Materials (*Towards a Natural Materials Policy*, 1972) it is predicted that by AD 2000 all the existing zinc mines in the USA will be exhausted.

2.4 MAIN OUTLETS FOR ZINC

Zinc is not a glamorous metal. In the pure state its mechanical properties are mediocre. It has no modulus of elasticity. It has a high density, which is three times that of aluminium and approaches that of iron. Alloyed with (60–70 per cent) copper it forms the brass series of alloys which have considerable industrial importance. It provides a useful series itself, when alloyed with 4 per cent aluminium, forming a series of alloys which have a high fluidity and a relatively low melting point, and are widely used for die casting. Zinc in the rolled form is used on the continent of Europe as a roofing material, in spite of poor creep-resistance. The outstanding characteristic of zinc, and probably its most valuable property, is its high resistance to atmospheric corrosion. Coatings can be applied to steel, either by hot dipping or electrolytically, which give valuable and often essential protection to the underlying metal. To a lesser degree, sprayed coatings of zinc and paint largely compounded of metallic zinc dust are used for the same purpose.

   Thus the main fields of application for zinc are:

Corrosion protection – galvanising
Zinc sheet
Die-casting alloys, containing 4% Al, 0.1% Mg and sometimes 1% Cu
Brass alloys

The consumption of zinc in these fields in 1982 for some of the main industrial countries is shown in Table 5.

Table 5

Pattern of zinc consumption in various industrial countries in 1982 ('000 tonnes)

	UK		USA		Japan		W. Germany		France		Italy	
	Wt	%	Wt	%	Wt	%	Wt	%	Wt	%	Wt	%
Galvanising	83.5	35	310	44	413	57	147	37	103	27	89	30
Die casting	38	16	158	23	109	15	89	22	46	12	53	18
Brass	59	25	73	11	101	15	90	22	67	18	111	38
Rolled zinc	14	6	37	6	29	4	66	17	99	26	13	4
Total	241	100	698	100	727	100	401	100	378	100	295	100

Source: World Bureau of Metal Statistics.

It will be noted that the pattern of consumption varies considerably. Consumers tend to be conservative and loth to move away from materials on which their practice has been established.

Demand and availability of zinc in the future

Table 6 shows how the world's zinc consumption has varied over recent years.

Table 6

World slab zinc consumption ('000 tonnes)

1950	1960	1970	1971	1972	1973	1974	1975	1976	1977	1978	1979	1980	1981	1982	1983
2076	3152	5097	5722	5555	6264	5971	5028	5754	5773	6209	6325	6131	6003	5957	6062

Figures from World Bureau of Metal Statistics.

Table 6 also shows that world consumption of zinc increased annually in a fairly regular way up to 1974, but since then the demand has tended to remain static. Figure 7 shows this graphically, and the figures for aluminium, copper and lead are included for comparison. The pattern of demand has been roughly similar, and the vicissitudes which the zinc market has suffered during the years of depression have also affected the other metals.

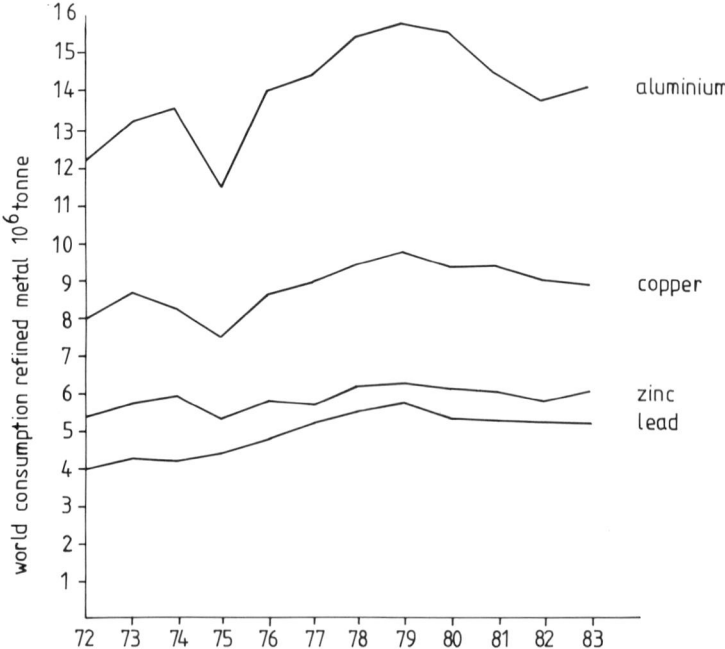

Fig. 7 – World consumption zinc, lead, copper, aluminium annual consumption since 1972. Figures from World Bureau Metal Statistics.)

The large increase in the cost of energy since the oil producers imposed the first rise in 1973 has had a profound effect on the world's economy since that date, and triggered off a trade recession which was still in being in 1983. Between 1972 and 1982 the price of oil increased from $2–3 per barrel to $39, and the cost of other forms of energy has risen proportionately. This has affected the zinc industry directly, in that efficient energy utilisation has now become of paramount importance, and the energy-consuming horizontal/vertical retort and the electrothermic processes find it extremely difficult to compete with the electrolytic or blast furnace operations, and most units have now been shut down.

The increase in energy costs since 1973 has also influenced the pattern of zinc consumption. One major outlet which has been affected is the die-casting field. Traditionally the automobile companies have consumed about 50 per cent of the zinc-base die castings produced. In 1968 the average American car carried 80 lb of zinc die castings, but under the pressure to reduce weight and improve fuel economy, the weight of die castings on each car had been reduced to 28½ lb [2]. This represents a serious reduction in the demand for the metal, but zinc die castings satisfy many outlets in others — namely in domestic hardware. Ever since the alloys were developed by the New Jersey Zinc Co. in the early 1930s, replacement by plastics has always been threatened. Although the number of plastics has increased and their properties improved, the zinc die-casting alloys have survived and still satisfy many applications, and there seems little reason to doubt that they will continue to do so. Again, in the field of corrosion protection, which is of major importance, substitutes have threatened from time to time, but for the corrosion protection of steel, zinc is still pre-eminent and likely to remain so. The same holds for the other important markets for zinc — in brass, zinc sheet and zinc crude pigments the overall market is unlikely to expand, unless the investigational and promotional efforts made by the International Lead Zinc Research Association (based in New York) and by a number of companies has a major success, but there seems little reason to expect demand to drop below its present levels. If international tensions can be contained, and the political and economic status quo continues, the present world demand for zinc of some 6 million tonnes per year should persist.

2.5 FUTURE OF ZINC CONCENTRATE SUPPLY

Most estimates of ore reserves must be speculative but one of the most authoritative has been given by the US Geological Survey [3] which estimates that the known resources of zinc which can be recovered with existing technology and economic conditions contain some 235 million tonnes of recoverable zinc.

In addition to the possible 235 million tonnes of recoverable zinc in 'identified' resources, the US Geological Survey estimates that a further 345 million tonnes of zinc may be contained in as yet undiscovered sources, of similar type and zinc content to those worked at present. Accepting these estimates — although the figure for undiscovered reserves must be subject to a large margin of error — a total of 580 million tonnes of zinc could be available for exploitation with existing technology and economic conditions.

In addition to the reserves which it is considered could be mined and processed to give conventional high grade zinc concentrates by existing techniques, the US Geological Survey also points out that there are large quantities of zinc existing in lower grade deposits, which under present conditions cannot be mined and recovered economically. The most important of these are the Kupferschafer and related beds in the Permian Zechstein formation, which is reported to be detectable across Europe from North East England to Silesia, and to be mineralised over a wide area. In Mansfeld, where it is mined for its copper content, it contains copper 2–3 per cent, zinc 1.3 per cent and lead 1.5 per cent, but in general it contains more zinc and lead than copper. There is evidence that large areas exist containing approximately 1 per cent as sphalerite. Richter [4] estimates that in Central Germany alone the zinc content of such deposits can amount to 250 million tonnes.

There are other low grade deposits the treatment of which is at present uneconomic, but which could possibly be worked in the future, such as occur in the Mississippi Valley and the Appalachian region of the United States. These are mineralised bodies in carbonate rocks of the same type as those already mined, but containing 1 per cent of zinc or less. The total resource potential of such deposits has been estimated at several hundred million tonnes of zinc. The manganese nodules on the sea bed of the Pacific Basin are stated to have an average content of 0.05 per cent zinc and 0.1 per cent lead. The problem of recovering these nodules is being actively studied and, if suitable techniques are developed, considerable quantities of zinc and lead could be produced.

It would appear therefore that there should be sufficient concentrates available to satisfy the industry well into the twentieth century. Its future would appear to be threatened more from the pressures building up from environmental control legislation, as is described in Chapter 9.

Prospecting techniques

The belief that, at least in the immediate future, the demand for zinc will be satisfied by the discovery of sufficient new ore bodies is encouraged by the recent appearance of a number of new factors. Knowledge of the geochemistry of zinc and the probable mode of formation of the main type of deposit has improved greatly in recent years. The development of the theory of plate tectonics has had a profound effect, and it is now possible to define with some precision the favourable geological environment likely to contain workable ore deposits. Geophysical techniques can be used, and aerial surveys have been developed which are based on the difference between the magnetic, electrical or gravimetric properties of ore bodies and those of the rocks which surround them, and these have been successfully used to detect anomalies suitable for more detailed examination.

Geochemical methods use the sensitivity and rapidity of the methods of chemical analysis now available for the estimation of small quantities of zinc and other metals in samples of rocks, soil, and certain types of vegetation. They enable an abnormal rise to be detected, and can assist in the justification of a drilling programme.

In the short term, it would seem likely therefore, that the mining and smelting industry should be able to satisfy the demand for zinc.

Recovery of zinc from scrap

As described in Chapter 4, the recycling of some zinc from scrap or discarded material is already practised, and whilst the amount of metal so returned could be increased, the effect on the overall zinc balance from even major improvements in recovery is unlikely to be great. Of the metal consumed per year approximately half is used in the galvanising, plating, and spraying of steel, and, on dismantling, only negligible quantities of the zinc can be recovered. In the brass industry the recovery of much of the available scrap is already complete since the value of the copper present in such scrap is high. Improvement in the recovery of scrap die castings could undoubtedly be made but this is unlikely to make more than a minor contribution to overall conservation of the metal. Some recovery of the zinc oxide present in used automobile tyres and other forms of rubber is already practised to some extent.

Thus, whilst improvement in conservation of scrap material should certainly be made, any effect on the zinc supply position would be marginal.

2.6 DETERMINATION OF ZINC PRICE

The London Metal Exchange (LME)

For many years, the price of zinc outside North America was based on the London Metal Exchange quotations, which can fluctuate from day to day according to the market situation.

The London Metal Exchange Company was formed in 1876 to act as a formal controlling body for metal trading, and it now acts both as a physical market, where warehouse warrants for fixed tonnages of accepted brands can be traded for immediate delivery, and also for transactions to be completed at a future specific date. Its importance has been considerably reduced by the introduction of the Producer Prices (see later).

There is a specific time for trading in zinc on the Exchange and numerous registered brands are accepted, the reference grade being GOB (Good Ordinary Brand) or Grade 4 zinc. There is trading at morning and afternoon sessions and the official price for the day is the market price at the close of the morning trading. A feature of the Exchange is that metal can be traded for any date up to three months forward, which enables both producers and consumers to hedge their commitment to purchase or sell metal up to three months ahead.

The prices for the sale of domestically produced zinc in the United States are set by the individual United States producers and their prices are generally closely in line. Domestic sales by Canadian producers tend to follow the United States prices.

There are differences from country to country in the way the cost of delivery and import duties is charged, but prices quoted to consumers usually cover delivery and duties, and are therefore higher than the published quotations on which they are based. Furthermore, premiums are charged according to the grade of zinc, the basic quotation usually being for G.O.B. or Prime Western (Grade 4) zinc. However, recently United States producers have based their selling price on a higher grade of metal, reflecting the trend for much of the new production to be purer metal from electrolytic plants.

The LME and Producer Prices
The zinc market has been bedevilled in recent years by currency realignments, price controls and an unsettled economic climate in most industrialised countries. At times the London Metal Exchange price has been unduly influenced by speculation. As a result of these and other factors the official price of zinc has tended to fluctuate considerably. During 1962, for instance, the price fell by nearly 10 per cent, whilst in 1963 it rose by over 40 per cent. Such instability was disturbing for both consumers and producers, and in 1964 the main European and Commonwealth smelting and mining companies decided to abandon the London Metal Exchange quotation and to quote their own price, known as the European Producer Price. Since 1964 most sales of zinc outside North America have been based on this Producer Price. Marginal supplies, mostly those imported from socialist countries, still continue to be based on the London Metal Exchange price.

The US Producer Price, which is published in *Metals Week,* is based on High Grade Zinc and is a weighted average of prices charged by individual North American producers. It is expressed as cents per lb.

The New York Metal Exchange was founded in the early 1890's, but United States merchants continued to do their business mainly through the London Metal Exchange for many years, with the result that the New York and London quotations tend to be closely related. The New York Exchange never became the established trading centre for zinc to the same extent as the London Exchange in the United Kingdom, and many producers entered into individual contracts with customers without the formalised barter of exchange dealings. Each producer quotes his individual price for his metal, but price quotations tend to be similar, and adjustments by one producer are usually quickly followed by others.

It will be seen (Table 7) that in general the prices move fairly closely together, but there have been a number of periods when they diverged sharply.

Table 7

Zinc price 1972–1983

Year	LME £ per tonne[a]	European Producer Price		US price (cents per lb[b])	
		$ per tonne[a]	£ per tonne on exchange rate[a]	Actual	Based on constant 1981 $
1972	151.0	390.75	156.3	17.75	34.38
1973	345.5	527.9	214.3	20.63	37.86
1974	528.4	777.5	332.3	35.95	60.59
1975	335.7	813.2	366.1	38.96	60.10
1976	395.0	795.0	440.4	37.01	54.26
1977	338.1	719.7	412.4	34.39	47.64
1978	309.1	607.3	316.6	30.97	39.98
1979	349.9	792.9	373.7	37.30	44.39
1980	327.4	798.6	343.3	37.43	40.85
1981	425.1	914.0	451,6	44.56	44.56
1982	425.5	846.6	483.8	38.47	36.29
1983	505.0	824.2	542.2	41.30	38.2

[a] World Bureau Metal Statistics.
[b] *Mineral Commodity Profiles Zinc, 1983.* US Bureau of Mines.

In 1971, as part of the US President's economic stabilisation programme, the price of zinc was frozen at 17 cents per lb, and controls were not entirely removed until December 1973. This had a serious effect on the US zinc industry and a number of plants were shut down. In 1973 shortage of supply caused the LME price to rise dramatically. In 1978 prices fell, owing to over-production coupled with reduced demand, and, as with other basic industries, depressed conditions persisted until 1983.

In the last column of Table 7 the US Producer Price (cents per lb) is calculated in terms of the 1981 $. It will be seen that even after eliminating the effect of inflation the real price of the metal has varied widely. In a number of periods the industry has been in a state of near penury. From Table 4 it will be seen that on the whole the overall demand for zinc has been maintained. Trouble has lain mainly in the fact that protracted periods of over-production have occurred, depressing the price. Because of the structure of the industry it is slow to react to reductions in demand. Contributing to this inflexibility is the time taken to bring new mines and smelters into production. Even though a prospect has been found and sufficient reserves proved, the mine and mill can rarely be brought into full operation within at least five years. Two to three years are required to erect a large extraction plant. Thus, in periods of high demand new capacity cannot be brought into being at short notice. Again, in periods of over-production the industry is reluctant to shut down capacity. Mines generally occur in remote locations, and miners, once dismissed and dispersed, cannot readily be re-assembled. In the case of most smelters, a reduction in output reduces profit-ability, and there is thus a great incentive to maintain production, and periods of over-supply are extended.

REFERENCES

[1] Fleming, M. G., *Transactions of the Institution of Mining and Metallurgy*, Vol. 82, Section A, p. 29, 1973.
[2] Broadhead, J. L., *World Symposium on Metallurgy of Lead and Zinc*, American Institute of Mining and Metallurgy, p. 20 1980.
[3] US Geological Survey Paper 820, United States Mineral Resources.
[4] Richter, G., Geologische Gesetzmassigkeiten in der Metallfuhrung des Kupferschiefers, *Archiv für Lagerstälten Forschung, Berlin, H.* 73, 1941.

3

Sintering and roasting

When William Champion first produced zinc in Europe at Warmley near Bristol in 1746, he used calamine as raw material, mined in the nearby Mendips. This material was then relatively plentiful and, as a pretreatment, only required heating to 600°C to drive off carbon dioxide, leaving zinc oxide, which could be treated in his crucible-type retorts.

As an ore frequently associated with calamine, the sulphide, variously called Black Jack, zinc blende or sphalerite, was already known, but this was a relatively inert material and would not produce zinc when heated with carbon. As early as 1758 William's elder brother John Champion obtained a British patent, No. 726, which covered the use of a roasting stage to convert Black Jack blende into zinc oxide, which could then be used in the distillation process. However, little use was made of this important discovery until the beginning of the twentieth century. Up to this time calamine was plentiful, and to roast blende with the primitive furnaces then available was a toxic, unpleasant operation. During this period, although zinc blende was recovered in considerable quantities associated with galena during lead mining, no use was made of it and it was discarded.

As the supply of good grade calamine began to be exhausted, attention was redirected to the use of blende, and the necessary roasting technique began to be developed. The birth of the flotation process at the beginning of this century – probably the greatest single development in non-ferrous metallurgy – with its efficient recovery of high grade sulphide minerals, ensured that in the future the zinc industry should be based on the sulphide, and today over 95 per cent of the world's zinc is produced from flotation-recovered blende. Today, with the exception of the pressure leaching method developed by Sherrit Gordon and described in Chapter 6 the first stage in all commercial extraction processes is the conversion, by roasting, of the sulphide into the more active zinc oxide. For the thermal processes, which require a granular product, some form of sintering is used almost exclusively.

3.1 THE ROASTING OPERATION

The main difficulty to be overcome in roasting is to ensure adequate access of oxygen to the surface of the solid blende particles, with rapid removal of the resulting sulphur dioxide gas. Since the blende is normally a flotation concentrate of high density and fine state of division, this is not easy. Sintering solves this problem because, in a well sized and conditioned mix, the sulphide particles are held in layers round the larger returns, and although they remain static, the high velocity and turbulence of the air passing through the bed enables rapid and complete reaction to occur.

Zinc sulphide is a relatively inert material, as it is not readily attacked by acids or alkalis and is unaffected when heated with carbon. In the case of both copper and lead, reaction between the sulphide and oxide takes place readily, and plays an important part in the smelting of these metals, but with zinc, as can be shown thermodynamically, the reaction

$$ZnS + 2ZnO = 3Zn + SO_2$$

does not take place continuously until temperatures in excess of $1300°C$ are reached. For these reasons the first stage in the recovery of zinc by all commercial processes is the conversion of the sulphide into the more reactive oxide.

The basic reaction can be summarised as

$$ZnS + 1\frac{1}{2}O_2 = ZnO + SO_2 + 431 \text{ kJ at } 1200°K$$

Although a considerable amount of heat is evolved, the roasting operation presents difficulties. Relatively high temperatures, in excess of $700°C$, must be maintained if the blende is roasted in air, and since the blende used is almost invariably a flotation concentrate, in a fine state of division and of high density, continuous agitation is necessary in order to obtain adequate gas—solid contact. Since use must be made of the evolved sulphur dioxide for the production of sulphuric acid, the sulphur dioxide content of the gases leaving the roasting furnace must not be allowed to drop below 4.5 per cent, or difficulties in the acid plant arise, and thus the use of excess air in the roasting operation must be severely limited.

These stringent requirements, and the necessity to reduce labour to a minimum, led to considerable evolution of the furnaces used. One of the most important of the early furnaces was of the Delplace type which consisted of 6—8 vertically superimposed hearths. The blende was charged into the top hearth, and then raked manually along the hearths to drop through a hole at the end into the hearth below, where the operation was repeated. The bottom hearth was heated from below, since in spite of the considerable amount of heat evolved during the combustion of the sulphide, the heat loss from the furnace was such that the temperature had to be augmented during the final stages. With good operation it was possible to reduce the sulphur content of the product to 1.5 per cent and to produce a gas containing 6 per cent sulphur dioxide, but the operation was manually exhausting and unpleasant. In consequence, considerable ingenuity was exerted on mechanising the operation.

The Ropp furnace, at one time widely used, employed rakes carried on endless chains driven mechanically, but the most successful solution was the adaptation of circular multiple hearth roasters of the Wedge or Herreshoff types, which were first used for roasting pyrites. These had a central shaft to which rabbles were attached which raked the ore alternatively across each hearth to fall into the one below, either at the centre or the periphery. With most furnaces it was necessary to burn some fuel in the lower hearths in order to complete the roasting operation, but with well-insulated large furnaces it was possible to work autogenously, and no extra heat was required.

Although these furnaces played a useful part and were widely used, they have now been entirely replaced. It was difficult to avoid build-up of accretions on the hearths and rabbles, and maintenance costs were high.

3.2 SINTERING

The mechanical hearth furnaces of the Herreshoff type were efficient in their way, but they had serious limitations. Even with a well designed and maintained rabbling mechanism, conditions for gas—solid contact essential for the roasting reaction were poor. A new principle was introduced in 1896 by Huntington and Heberlein at Pertusola in Italy by the development of blast roasting. Whilst used initially for roasting galena, it was found that the method could also treat blende. Steel crucibles holding about 1 tonne of charge were used, and air was blown under pressure through a grid at the bottom of the crucible. The operation was started by igniting a layer of coal on the grid, and then feeding in concentrates, which ignited and burnt rapidly under the influence of the blast. The blast rate was controlled so that the temperature reached was sufficient to partially melt the ore particles, so that they formed a biscuity porous cake. Further additions of concentrate were then progressively made until the crucible was full, and when roasting was complete, the crucible was emptied and the process repeated.

Although the pot roasting method was important as it introduced the blast roasting principle, it was not widely used, being almost immediately displaced by the development of Dwight-Lloyd sintering, which mechanised the method and made continuous operation possible. A Dwight-Lloyd machine is similar in construction to the chain grate stoker used for coal-fired boilers. It consists essentially of a series of flat pallets fitted with bars forming a grate on which the charge rests. These travel along rails over a series of windboxes, and at the end of the machine they drop around curved guides to travel back under the windboxes to the feed end of the machine, where they are picked up by a large, electrically driven, sprocket wheel which replaces them on the rails over the windboxes and provides the motive power for the operation.

The charge to be roasted, suitably proportioned and conditioned, is fed from a bin above the machine, and a layer of predetermined height (generally

15—18 cm) is laid on the slowly moving pallets. After receiving its charge each pallet passes immediately over the first windbox, which being under suction begins to draw air down through the charge. At the same time it passes under the igniter, which is a refractory lined box covering the width of the pallet and fired to a high temperature by gas or oil. The hot combustion products are drawn on to the bed and ignite the top layer of charge. The pallet then continues along the machine and the windbox suction below continues to draw air down through the bed, causing a thin layer of combustion to pass down through the charge. The narrowness of this combustion layer or zone is an essential feature of the process. Under normal sintering conditions the reaction zone moves down through the bed at approximately 0.25 mm/sec and the velocity of the gas through the bed is approximately 1000 times greater. The intensity of reaction is indicated by the fact that, with blende sintering, temperatures in excess of 1450°C can be recorded momentarily by thermocouples placed in the bed.

Under good conditions, this zone takes about 20 minutes to pass through the bed and complete the roasting operation. The speed of the machine and the air volume sucked through the bed are adjusted so that completion occurs by the time the last windbox is reached. The roasted cake, the sulphur content of which has been reduced to under 1 per cent, then falls off the pallet, as it reaches the end of the machine, into a bin from which it is withdrawn for crushing and subsequent treatment. The lower part of the cake is generally at a dull red heat at this stage (although the top is cold), and a cooling table or drum is frequently used to prevent fume emission and facilitate subsequent handling.

Conditions essential for down-draught blende sintering

Zinc blende sintering is not an easy operation and a number of conditions must be satisfied for success to be attained — there can be few metallurgical processes where detailed attention to basic details such as sizing, proportioning and mixing is so essential. The process depends entirely on obtaining uniform ignition and even development of the narrow intense combustion zone which passes down the depth of the bed. Unless conditions across the bed are uniform, uneven roasting will take place, blow holes will develop and roasting will be incomplete.

Sulphur content of charge.

The first essential for correct hot zone conditions is the attainment of the optimum sulphide content of the feed, since this provides the fuel for the reaction. If insufficient sulphide sulphur is present the reaction will be completed only partially, but if sulphur is in excess, the hot zone will fuse and lose porosity, roasting again being imperfect. For down-draught blende sintering with normal concentrates, the sulphide content of the feed should be between 5.5 and 6.5 per cent with 6 per cent as an optimum. Since the sulphur content of most blendes is of the order of 30 per cent, satisfactory sinter cannot be made from a pure blende charge. In the early days this difficulty was overcome by preroasting the blende in a hearth furnace to reduce the sulphur content to 6–8 per cent. In 1928, S. Robson of the National Smelting Company at Avonmouth proposed an alternative solution, which involved taking five-sixths of the output from the machine, crushing it, and using it as inert material (or returns) to dilute the sulphur content of the incoming blende so that satisfactory sintering could occur. This meant that on average each particle of sinter passed five times over the machine and, although this involved considerable handling of materials, it worked well and has been adopted by most plants requiring soft blende sinter.

Sizing of returns
Adequate porosity of the bed is essential for good sintering and the flow of air down through it must be uniform. The main factor in controlling porosity is the sizing of the returns, since these are the major component of the feed. They are obtained from the machine discharge after cooling by screening through a 4 mm screen, the remaining 20 per cent of the discharge passing forward to output.

Moisture content
The moisture content of the feed is also a critical factor, since it is essential for ensuring physical contact between the blende and return particles. A dry mix will not sinter satisfactorily, but porosity is lost if the mix is too wet, and again sintering cannot occur. With well-sized materials the moisture content is held within the range 6–7 per cent, and to do this automatic control is widely used.

Proportioning of charge
Great care is taken to proportion the charge accurately. Most plants treat more than one blende simultaneously, together with oxides and other materials. These, together with the returns, are held in separate bins discharged by controlled weight feeders, so that the final sulphur content and composition of the mix can be held within close limits.

Mixing
This is an important operation, as it is obviously essential that during mixing the blende which forms the fuel should be spread as uniformly as possible around the coarser return particles. Paddle or drum mixers are used for this operation.

Feeding the machine
After the charge has been proportioned, moistened and mixed it must be fed to the machine with great care. Any unevenness of pressure causes serious variations in porosity and therefore of air flow. Special mechanisms such as swinging chutes or belts which traverse the width of the pallets are used to give even deposition.

Gas handling
An essential requirement of the sintering operation from the standpoint of both hygiene and economics is that all the sulphur dioxide liberated during roasting

should be collected and converted into sulphuric acid. Thus, whilst all the evolved gas must be recovered and since contact sulphuric acid plants have difficulty in treating economically gas containing less than 4.5 per cent sulphur dioxide, little leakage of free air into the system can be permitted. This is yet another reason why it is essential that the sintering zone pass evenly down the bed. No blow holes can be tolerated, and suction must be cut off as soon as the hot zone has traversed the cake and roasting is completed. As the pallets pass along the machine the sulphur dioxide content of the gas drawn through them varies. In Fig. 8 a typical distribution of gas strength, for a ten-windbox machine, obtained by withdrawing samples from the windboxes, is shown. The curves shows that from the pallets in the middle of the machine gas strengths of from 10 to 12 per cent sulphur dioxide can be obtained, which approaches the theoretical maximum of 15 per cent.

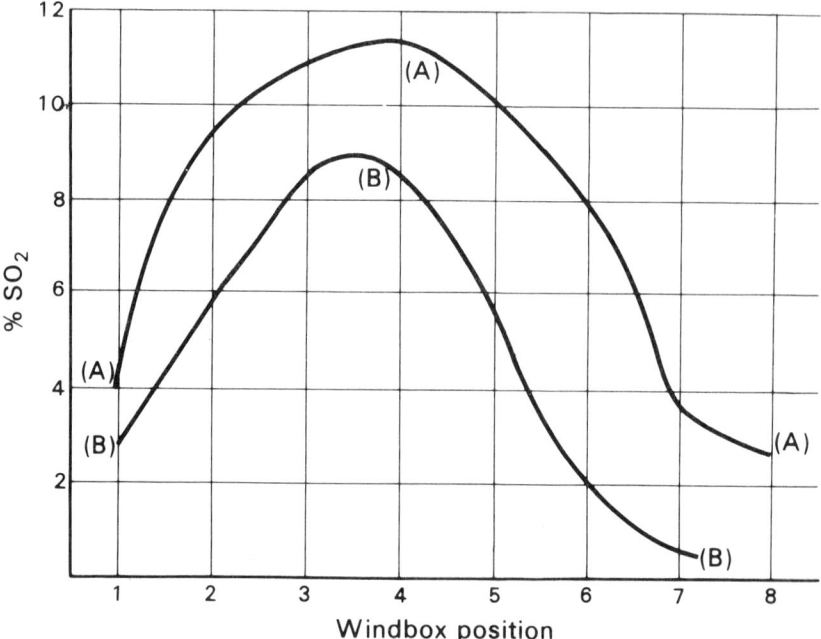

Fig. 8 — Distribution of sulphur dioxide on down-draught sintering machine.
A: Soft (low lead) down-draught. B: Hard (high lead) up-draught.

Gas recirculation

Most sintering machines employ some form of gas recirculation to enrich the gas passing to the acid plant. With a ten-windbox machine the gas from the first five windboxes passes to the acid plant, that from the remaining five being drawn off by an auxiliary fan and fed into a hood covering the first 75 per cent of the bed.

It is then drawn down through the pallets in this front section both to act as combustion air and to become enriched in the process. Using such a scheme a typical figure for the sulphur dioxide content of the circulating gas would be 2.5 per cent and the tenor of the gas passed to the acid plant would be 7 per cent sulphur dioxide. Gas circulation therefore performs a useful function in improving the economics of the acid plant.

Elimination of cadmium and lead

Most blendes contain from 0.1 to 0.5 per cent cadmium and this represents a valuable byproduct in zinc smelting. Cadmium compounds are more volatile than those of zinc, and in the down-draught sintering operation approximately 70 per cent of the cadmium is volatilised, largely as cadmium sulphide. Most of this is condensed and collected on the pallet bars, and is recovered by knocking the bars with revolving hammers or chains, as the pallets return underneath the machine. The displaced material, containing from 4 to 6 per cent cadmium, is collected and the cadmium recovered by leaching with sulphuric acid in a separate plant. Approximately 20 per cent of the lead present in the blende is eliminated by volatilisation during soft sintering (roasting of low-lead-containing zinc blende), and is collected in the acid plant purification system as a lead sulphate sludge, which is sold.

Operating data

Table 8 gives typical operating data for a blende sintering machine producing sinter for a vertical retort plant.

Table 8

Operating data for a zinc blende sintering machine

Pallets	length — 1 m, draughted width — 2 m
Number of pallets draughted to windboxes	21
Number of windboxes	10 + 1 igniter box
Machine speed	0.75–1.0 m/min
Bed suction	3–4 kPa
Bed depth	18 cm
Moisture content of mix	6–7%
Sulphur content of mix	5.5–6.5%
Gas volume at acid plant	283 m^3/min
SO_2 content of gas at sinter plant	7–7.5%
Sinter output	114 tonnes/day
Sulphur eliminated per m^2 grate area/day	1 tonne
Sinter analysis	Zn 61–64%
	Pb 0.8–1.8%
	Fe 8–10%
	S 0.3–0.7%

Up-draught zinc—lead sintering

Down-draught blende sintering was developed to produce a roasted granular product as a suitable feed for the horizontal or vertical retort processes. For this purpose it was very satisfactory and plants operating these processes today use it as the main desulphurising stage.

With the development of the zinc—lead blast furnace a number of additional requirements had to be met. In the first case, the furnace required a much more lumpy product than the retort processes, with the sizing of the charge for a blast furnace between 2.5 and 10 cm. Any material below 1 cm is rejected. The presence of lead in the blast furnace sinter also caused complications since, with down-draught sintering, lead compounds tend to migrate through the bed and collect on the pallet bars, causing excessive scaling and corrosive attack.

The orthodox lead smelters had also experienced the same difficulties, although the problem was less acute, since in lead sulphide sintering the temperatures reached are appreciably less than in blende roasting. A revolutionary concept was developed at Port Pire in Australia and by Stolberg in Germany which largely solved the problem. This involved reversing the flow of air and blowing upwards from below the bed. The success of this work in lead sintering encouraged trials of the new method at Avonmouth. After initial difficulties these were successful, and up-draught sintering is now used by all plants operating zinc—lead blast furnaces. A typical updraught zinc—lead sintering machine is shown diagrammatically in Fig. 9.

The main differences introduced by the new method, by comparison with soft sintering, are:

Ignition

It is still necessary to ignite the charge by down-draughting and for this, sufficient charge to form a 2.5 cm layer on the bed is fed ahead of the igniter box. This layer is kindled under the charge, then passes over a dead phase 60 cm long. The main charge is added, to a depth of 30 cm, and when the first windbox is reached, air is blown up from below and roasting upwards through the charge commences.

Gas handling

Since the air flow through the bed is upwards instead of down, a hood must run the whole length of the machine in order to collect the gas for treatment in the acid plant, as shown diagrammatically in Fig. 9. In order to obtain the maximum possible sulphur dioxide content of the final gas, the primary fan blows air into the end windboxes only. The gas collected in the hood above these boxes is then withdrawn by the main circulating fan and passed into the early boxes, so that enrichment occurs. The weak wet gas leaving the bed above the first windbox is recirculated through a lagged main to the intermediate fan inlet.

Feeding the machine

Feeding an updraught machine is complicated by the fact that the feed must be split in order to provide the ignition layer (2.5—3 cm thick) before the igniter

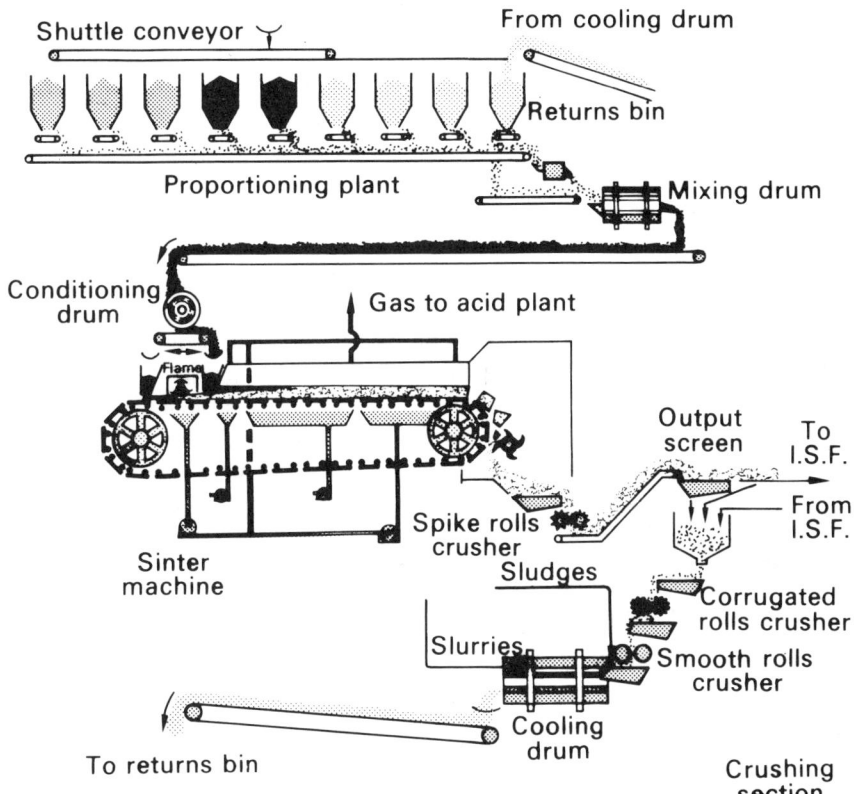

Fig. 9 – Typical flowsheet for up-draught sinter machine.

box, the main feed being added immediately afterwards to raise the total bed depth to 25–30 cm. Since with this type of sintering the rising gases tend to lift the charge from the bed, it is again most important that the mix should be correctly sized, well mixed and conditioned. Great care must be taken to ensure it is evenly fed on the machine.

The sintered output

The sintered cake leaving an updraught zinc–lead machine, since it is to form the feed to a blast furnace, is made much harder than in soft sintering. The hot cake drops from the end of the machine into a pronged breaker which breaks it into pieces below 15 cm. It then passes to a cooling table or drum, similar to that used in soft sintering. The cooled sinter is screened over a 2 cm screen, the oversize passing forward to output, whilst the undersize is crushed in cone type crushers until it all passes through a 4 mm screen. It is then added as returns to reduce the sulphur content of the incoming zinc and lead sulphides until a value of approximately 6.5 per cent sulphur in the feed to the machine is reached.

A number of papers can be consulted for more details of the up-draught sintering process [1] –[6] .

The horizontal and vertical retort and electrothermic furnace operators, who have to use good grade concentrates, use down-draught machines, but the blast furnace plants are forced to use up-draught machines because of the high lead content of their charge.

Thus sintering, as practised today, offers an almost complete solution to the physical problems involved in roasting. Since both the horizontal and vertical retort processes require a granulated charge, sintering is used almost exclusively as a pretreatment. For the production of a hard lump feed for the blast furnace, sintering is essential at the moment, although hot-briquetting techniques now under development may offer a viable alternative. From an operating point of view, however, sintering is expensive, as no recovery of heat is possible and, in fact, large quantities have to be dissipated. Capital and labour costs are relatively high, and since considerable tonnages of abrasive material have to be crushed and circulated, maintenance costs are considerable. The electrolytic process can use the much cheaper fluidised bed technique, thus obtaining an appreciable advantage.

3.3 ROASTING FOR THE ELECTROLYTIC PROCESS

In the electrolytic process, the roasted product is leached in sulphuric acid, and fine material is advantageous. For the roasting stage almost all electrolytic plants today use fluidised bed roasters, although in some of the older units suspension type roasters are still employed. Fluidised bed roasting has many advantages over sintering — very large single units can be built with low operating and maintenance costs. Whilst the fine state of division of the calcine from such furnaces is of great benefit in electrolytic practice it is unsuitable for thermal smelters, which must use sintering techniques to give the granularity required.

In the early days of the electrolytic process, the main aim during the roasting stage was to produce a calcine which gave maximum zinc but minimum iron solubility, because of the difficulty in filtering and washing free from zinc the ferric hydroxide precipitate which was then the standard method of iron removal. This was complicated because, when heated, zinc and iron oxides readily form a compound, zinc ferrite ($ZnO.Fe_2O_3$). If the iron occurs in the original concentrate as marmatite, in which it is present as a solid solution in the zinc sulphide, ferrite formation is almost complete. Even when present as discrete mineral particles such as pyrite, at temperatures above $900°C$, which are necessary for an adequate roasting rate to be maintained, over 90 per cent of the iron present forms ferrite, irrespective of the form in which it was originally present.

Whilst zinc ferrite is readily soluble in hot leaching acid — both the zinc and the iron being dissolved — it is but little attacked at lower temperatures. Today, with the development of modern pH control and such methods of precipitating iron as jarosite and similar compounds (see Chapter 6), problems due to iron have virtually disappeared, but when the electrolytic process was first developed this was not so, and leaching was carried out at temperatures at which ferrite was not attacked, and the zinc oxide associated with it was left in the residues. As a consequence, whenever possible, concentrates low in iron were used and the application of the process was limited.

Although the thermal methods of zinc production adopted sintering for roasting almost exclusively, it was little used for the electrolytic process. Thermodynamic studies indicate that at low oxygen pressure zinc ferrite is unstable, and free zinc and iron oxides are formed. High temperatures such as those reached during sintering also favour ferrite decomposition. Thus, whilst sintering produces little ferrite and a product with a high zinc solubility, most of the iron is also soluble and precipitation problems were serious. In addition, zinc silicate was formed which gave high concentrations of silica in solution, which again caused handling difficulties. In general therefore, sintering has had little application in electrolytic zinc production, and the industry has turned to other methods of roasting.

Suspension roasting

At first, roasting for the electrolytic process was carried out in large circular hearth furnaces of the Wedge or Herreshoff type. It has been pointed out, however, that these give only limited gas—solid contact and a major step forward was taken at Trail, in Canada, by the Consolidated Mining & Smelting Company with the development of suspension roasting. This method used the principle already developed for pulverised fuel firing. Dry, finely crushed blende was fed with air through a burner into a hot combustion space which was held at a temperature of about 950°C. The blende ignited and formed a flame in which roasting was rapidly completed. The first successful attempts used existing Wedge furnaces with the upper hearths and rabbles removed to form the combustion or roasting chamber. The two bottom hearths were retained and were used to dry the incoming blende, which was then crushed in a ball mill. The fine dry blende was fed with air through burners into the hot chamber above, and under these conditions sulphur could be almost completely eliminated and a product containing 0.8—0.5 per cent sulphur produced.

Approximately 40 per cent of the roasted product was removed in the gas stream, which contained 9—10 per cent sulphur dioxide at a temperature of 950°C. Handling this dusty hot gas posed special problems. It was passed first to a waste heat boiler designed to deal with high dust loads. About 1 kg of steam was generated in this boiler per kg of blende roasted. The gas leaving the boiler at a temperature of approximately 300°C passed to a cyclone where the remainder of the dust was collected, and the gas was ready to enter the acid plant purification system.

For some time, suspension roasters with their high capacity — single units roasting over 300 tonnes of blende per day were built — and low maintenance costs, displaced circular hearth furnaces entirely. Now, in their turn, they have been largely replaced in the electrolytic industry by the latest development in roasting technique, the fluidised bed process.

Fluidised bed roasting

The development of fluidised bed techniques in the mid-1930s was a major advance in technology, of which advantage was rapidly taken in both the chemical and metallurgical industries. The first large-scale commercial application was in Germany where it was used in the Winkler process for water gas generation from lignites, but soon afterwards the oil industry also adopted the system for catalytic cracking, and showed its advantages for gas–solid reactions in which fusion does not take place. In metallurgy, the first major application was in the roasting of flotation pyrites, and success in this field led to its use in zinc blende roasting.

The method depends upon the fact that if a current of gas at low velocity is passed up through a bed of solids it can flow through the interstices between the particles, which remain in contact with each other, with little movement. As the velocity of the gas is increased, however, the frictional forces exerted on the particles begin to exceed their weight, lifting occurs, and the particles become mobile within the bed, which begins to resemble a liquid of high viscosity. In this stage the onset of fluidisation has begun. When the gas velocity is increased further, the volume of the bed expands, movement of the particles within it becomes rapid, and the bed enters a highly turbulent state, known as dense-phase fluidisation. If the gas velocity is increased still further, particles begin to be carried upwards out of the bed, and the phase of pneumatic transport begins.

In the state of dense-phase fluidisation, a considerable amount of continuous circulation of both gas and solids takes place, and almost complete uniformity in composition is reached. Fresh solid added to the bed quickly reaches the bed temperature, as does the incoming gas, and the high relative velocity of the gas over the solid surfaces favours chemical reaction.

The conditions under which fluidised beds are obtained, and the range of gas velocities within which they operate, depend upon a variety of factors, the two most important being particle size and superficial gas velocity, but particle shape and density, and the viscosity and density of the gas have also considerable effect. An analysis of the relationship of these factors is given by Schytel [7].

The method is particularly applicable to the roasting of zinc blende, and for the electrolytic process which requires fine, easily leached material, it has displaced almost all other roasting systems. Most blende flotation concentrates produced today are relatively high in grade, containing little lead or other material likely to cause agglomeration in the bed, and consequently bed temperatures of $900°C$ and above can be maintained without loss of free flowing characteristics. At such temperatures the basic roasting reaction

$$ZnS + 1\tfrac{1}{2}O_2 = ZnO + SO_2$$

proceeds rapidly and completely; equilibrium lies far to the right and the reaction is virtually irreversible. The sulphide sulphur content can be reduced to less than 0.5 per cent in spite of the fact that the chemical driving force is reduced, since the concentration of fresh blende added to the bed is greatly diluted by the large excess of already roasted calcine, and the average concentration of oxygen is that of the exit gas at approximately 3—5 per cent — disadvantages which are more than compensated for by the high temperatures employed, and the excellence of the gas—solid contact. Large units can be built: single furnaces capable of roasting 500 tonnes of blende per day are in operation, and at Risdon in Tasmania a furnace has been built which will roast 900 tonnes of blende per day [8]. Fluosolid furnaces contain no moving parts internally, and maintenance and labour costs are low. High strength gas containing 10 per cent sulphur dioxide can be produced, and thus the size of the sulphuric acid plant required is minimised. The process is thermally efficient, and the waste heat boilers used to cool the hot gas leaving the cyclones can produce approximately 1 kg of high pressure steam per kg of blende roasted.

Fig. 10 – World's largest fluid bed zinc concentrate roaster at the Risdon, Tasmania, plant of the Electrolytic Zinc Company of Australasia Ltd.
(Courtesy of Electrolytic Zinc Company of Australasia.)

The process becomes increasingly difficult to operate if the grade of concentrate falls, for, with a lead content above 3 per cent, agglomeration begins to occur, interfering with the operation of the bed. The majority of blende concentrates, however, can be treated without difficulty and, as a pretreatment for the electrolytic process, fluidised bed roasting is an almost ideal solution. The fine state of division of the product precludes its use in the blast furnace process unless some form of briquetting or sintering is used, and the inability of the orthodox fluosolid roaster to treat lead-containing materials is a disadvantage. This is largely overcome by a system developed by Metallurgie Hoboken-Overpelt S.A. in Belgium. The furnace is rectangular and some 5.5 m high. The charge of concentrates and recycled fines is homogenised with sulphuric acid and then pelletised to 0.4–0.5 mm diameter. After drying, the pellets are fed to the roaster. Standard fluidisation conditions exist for 85 per cent of the length of the bed, and the charge then overflows into a finishing bin, where air is also added to complete oxidation. Relatively high bed temperatures of 1050–1060°C are used. Because the charge is pelletised, the dust carry-over to the waste-heat boiler is only of the order of 20 per cent. Up to 15 per cent of lead can be tolerated in the charge, which is a considerable advantage offset by the cost of pelletising.

The roasting capacity of fluosolid furnaces is high: most units operating the slinger belt method of feeding eliminate from 1.8 to 2.1 tonnes of sulphur per square metre of grate area per day, which is equal to that attained on many sinter machines. Furnaces using a slurried feed tend to run at approximately 70 per cent of this rate.

Some measure of the time of roasting in a fluidised bed furnace is indicated by a test reported by Canadian Electrolytic Zinc [9] using radioactive zinc concentrates which showed an average residence time in the bed for the furnace overflow material of 5 hours, and for the dust carried over of 1 hour. The time taken to renew the bed completely was approximately 20 hours.

Sulphur elimination

For reasons which have been discussed, the oxidation of zinc sulphide in a fluidised bed roaster is almost complete, and many plants produce calcine with a sulphide sulphur content as low as 0.1 per cent, the results falling mostly within the range 0.1–0.4 per cent. The sulphur present as sulphate in the calcine is always higher, varying from 0.8 to 2.5 per cent. The presence of some sulphate is desirable since it compensates for losses in the leaching and electrolysis stages, but if in excess, build-up in the cycle occurs and some discard is required.

Under normal conditions with temperatures above 900°C, zinc sulphate should not exist in the furnace, and a number of studies of the thermodynamic properties of zinc sulphate, basic zinc sulphate, and the system zinc–sulphur–oxygen have been made [10–13]. It can be calculated from this work that whilst, in the absence of sulphur dioxide, zinc sulphate decomposes at 725°C forming zinc oxide, in a roast gas containing from 6 to 12 per cent sulphur dioxide the basic sulphate, $2ZnSO_4.ZnO$, is stable up to 850°C, but not at higher temperatures.

As the dust-laden gas cools in the waste-heat boilers and cyclones, the possibility of sulphur trioxide formation increases and sulphation can occur. The production of sulphur trioxide is encouraged by the concentration of oxygen in the gas; the proportion of excess air is therefore carefully controlled and is rarely allowed to exceed 10–15 per cent. Under such conditions the proportion of sulphur in the sulphur trioxide caught in the acid plant purification system can be reduced to less than 1 per cent of the total roasted.

The furnace
Most turbulent bed roasters are cylindrical in shape and consist of a windbox into which air is blown under pressure, below a hearth through which penetrate a large number of alloy steel tuyeres to even distribution through the bed above. The grate area of the large furnace built at Risdon (capable of roasting 900 tonnes per day of blende) is 123 m^2. It is constructed of abrasive-resistant castings 150 mm thick, supported on beams and fitted at 100 mm spacing with 13,000 integrally cast high chrome alloy nozzles, 150 mm long with a 4 mm diameter air passage.

The height of the bed in all furnaces is controlled by the position of over-flow ports, which are usually 1–1.5 m above the hearth. The diameter of the furnace above the hearth is increased to give a freeboard section in which oxidation of most of the fine particles carried out from the bed can be completed. The total furnace height above the bed lies generally in the range 9–20 m.

Differences in practice arise, mainly in the methods used to feed the incoming blende into the furnace. In the system developed by the Dorr Company the blende is first screened and slurried with water to a consistency of 75–80 per cent solids, which is pumped into the furnace through a number of nozzles just above the bed surface, care being taken to obtain even distribution. An alternative system projects or slings the blende into the furnace by means of fast moving belts and, under such conditions, a large proportion of the finer particles do not reach the bed but are roasted whilst in suspension in the freeboard section.

Control of temperature is very important. In order to obtain high throughput and complete roasting, it is necessary to run as high a bed temperature as possible, but if it is too high, agglomeration begins and the bed no longer operates smoothly. Bed temperatures generally lie between 900°C and 980°C depending on the mineral composition of the blende treated. The oxidation of zinc sulphide is strongly exothermic, and the heat liberated is greater than that necessary to maintain these temperatures. Some form of cooling is essential, and this is frequently effected by direct additions of water to the bed. With the slurry system of feeding, most of the water required enters with the feed, a final adjustment being made through sprays within the furnace. Cooling coils, carrying high pressure water, embedded in the bed are sometimes used, but this system tends to be inflexible if a variety of blendes is treated, although again adjustments can be made by adding water through sprays.

Fluidisation conditions

The particle size of most flotation blendes varies from 50 to 300 μm, the mean particle diameter lying in the range 175–225 μm. With such material, dense-phase fluidisation can be obtained, with a void fraction varying from 0.6 to 0.8, and a space velocity lying between 30 and 45 cm per second. With an air volume approximately 10 per cent greater than that required stoichiometrically to roast all the sulphides present, satisfactory fluidisation and roasting can be obtained with most blendes, yielding a gas containing 8–10 per cent sulphur dioxide.

At most plants, gas velocities are used which are sufficient to carry the finer particles out of the bed into the freeboard space and ultimately out of the furnace, and only 30–50 per cent of the calcine is collected from the furnace overflow chutes. The dust-laden gas passes immediately into a waste-heat boiler where the temperature is reduced to 350°C. The boilers are almost invariably of the Lamont forced circulation type, and, since they collect a large proportion of the calcined blende, the dust loading is heavy and mechanical cleaning of the walls and tubes is essential. The dust is held in hoppers and removed continuously or periodically. The gas leaving the boiler is further cleaned in cyclones, and then finally in electrostatic precipitators, before passing to the sulphuric acid plant. The calcine is therefore collected from several locations, a typical distribution being as follows:

furnace overflow	40%
waste-heat boilers	20%
cyclones	35%
electrostatic precipitator	5%

The distribution may vary with plant practice depending on the relative fineness of the blendes treated and the gas space velocity employed. In some cases almost all the calcine may be carried out of the furnace in the gas and collected in the gas cleaning and cooling stages. A brief description of fluidised bed roasting carried out in eight electrolytic zinc plants is given in the literature [14].

REFERENCES AND FURTHER READING

[1] Woods, S. E. and Harris, C. F., Factors in zinc–lead sinter production, *Port Pirie Sintering Symposium,* Australian Institution of Mining and Metallurgy, 193–218, 1958.

[2] Sellwood, R. M., Updraught zinc–lead sintering, *Mining Journal,* Vol. 254, 15th April, pp. 434–435 1960.

[3] Green, A. D. M. and Andrews, B. S., Sintering developments at Cockle Creek, *Proceedings of Institution of Mining and Metallurgy,* Vol. 212, Dec., pp. 41–59, 1964.

[4] Woods, S. E. and Harris, C. F., Heat transfer in sinter roasting, *Symposium on Chemical Engineering in the Metallurgical Industries,* Institution of Chemical Engineers, pp. 77–86, 1963.

[5] Evans, C. J. G., Sintering techniques as applied to the Imperial Smelting Process, United Nations Industrial Development Organisation Meeting on lead and zinc industries, Report 1D/45, Paper No. 1D/WG 33/6, p. 51, 1970.

[6] Shoobridge, D. H. and Coppock, B. W., Improvements in sintering practice at Cockle Creek, *Australasian Institution of Mining and Metallurgy Conference, Newcastle, N.S.W.,* pp. 349–56, May 1972.

[7] Schytel, F., *Metallgesellschaft Review of Activities,* Vol. 1, p. 13, 1959.

[8] Lightfoot, R., Fluid bed roasting of zinc concentrate at Risdon Tasmania, *Australasian Institute of Mining and Metallurgy, Tasmania,* pp. 359–365, May 1977.

[9] Heino, K. H., McAndrew, R. T. and Ghata, N. E., Fluid bed roasting of zinc concentrate at Canadian Electrolytic Zinc Limited, Valleyfield, *American Institute of Mining and Metallurgical Engineers Symposium on Mining and Metallurgy of Lead and Zinc,* Vol. 2, p. 157, 1970.

[10] Kelley, K. K., The thermodynamic properties of sulphur and its inorganic compounds, Bulletin No. 406, US Bureau of Mines, 1937.

[11] Kellogg, H. H., Equilibrium considerations in the roasting of metallic sulphides, *Transactions of the American Institute of Mining and Metallurgical Engineers,* Vol. 206, 1956.

[12] Stephens, The fluidised bed sulphate roasting of non-ferrous metals, *Chemical Engineering Progress,* Sept., 1953.

[13] Fisher, Some applications of physical chemistry to the sulphate roasting of metallic sulphides, *Transactions of the Institution of Mining and Metallurgy,* Vol. 73, pp. 109–176, 1963–4.

[14] *American Institute of Mining and Metallurgical Engineers Symposium on Mining and Metallurgy of Lead and Zinc,* Vol. 2, pp. 128, 149, 183, 201, 227, 253, 277, 311, 1970.

4

Reduction of zinc oxide

Although zinc had been produced since the fourteenth century, or even earlier, by processes involving reduction and condensation, little was known of the physical chemistry involved, and the various methods used to carry out the operation were developed empirically.

4.1 THE PHYSICAL CHEMISTRY OF ZINC OXIDE REDUCTION

In 1917, Bodenstein showed that the reduction of zinc oxide by carbon

$$ZnO(s) + C(s) = Zn(g) + CO(g) \tag{i}$$

(the overall reaction occurring in both horizontal and vertical retort processes) proceeded by two successive gas–solid reactions which were readily reversible:

$$ZnO(s) + CO(g) = Zn(g) + CO_2(g) \tag{ii}$$

$$C(s) + CO_2(g) = 2CO(g) \tag{iii}$$

The first thorough thermodynamic study of the reduction was carried out by C. G. Maier and his colleagues at the US Bureau of Mines [1]. They collected together, and where necessary determined, all the relevant thermodynamic data, producing consistent values. From these they calculated the free energy of zinc oxide, and were able to specify the temperature at which continuous reduction of zinc oxide could begin, and the proportions of zinc vapour, carbon monoxide and carbon dioxide which would exist at all temperatures of interest in retort smelting. The work was a landmark in the metallurgy of zinc, and provided for the first time a theoretical background to the production of the metal.

The zinc product of the reduction process is always the vapour phase because the reduction of zinc oxide by carbon at normal pressures does not proceed significantly below the zinc boiling point of 907°C. The thermodynamic expression of this fact can be seen in Fig. 11, which plots the free energies of the relevant reactions as a function of temperature. In simple terms the oxide of carbon only becomes more stable than the oxide of zinc at temperatures above the boiling point of zinc. The figure also contains free energy plots of other metal oxides generally associated with zinc ores. The lower the line on the figure the more stable is the oxide, indicating that in retort smelting iron, lead and copper are also reduced but that refractory oxides such as silica, and the even more stable alumina and lime, remain unreduced in the gangue.

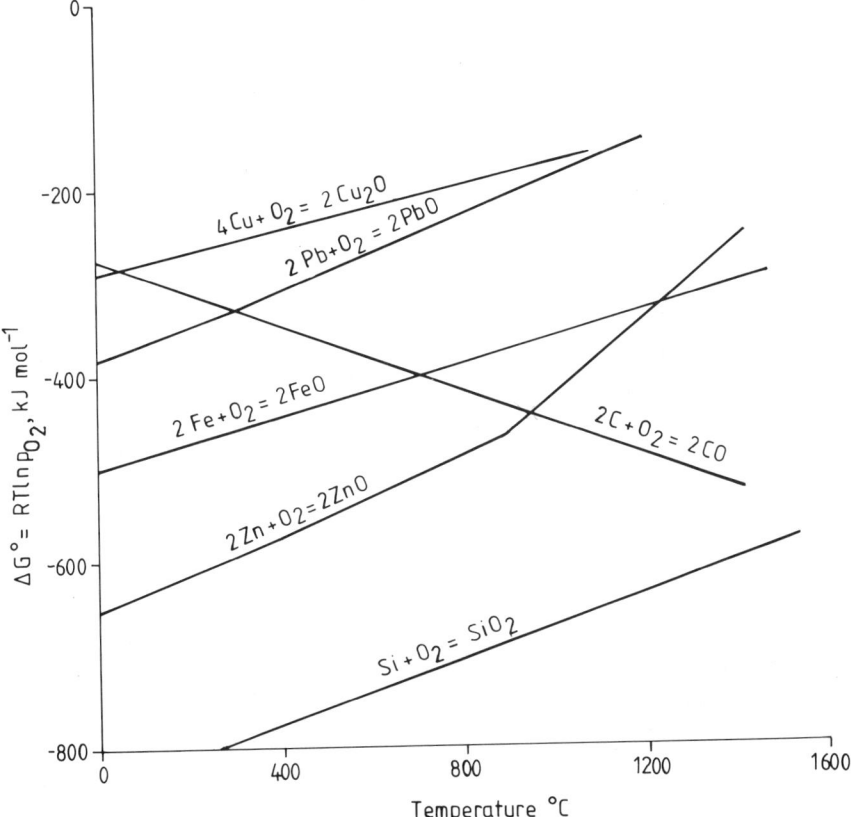

Fig. 11 — Standard free energy of formation of metal oxides as a function of temperature.

From equations (ii) and (iii) it is obvious that under retort conditions, as far as carbon dioxide is concerned, there are two opposing tendencies: reaction (ii) tending to increase, and (iii) to decrease the carbon dioxide content. The equilibrium conditions for the two reactions can be calculated from the tables of Barin and Knacke [2] at 1200 K.

Reaction (ii) $\Delta G^\circ = 183{,}685 - 115.6T \ \text{J mol}^{-1}$

Reaction (iii) $\Delta G^\circ = 169{,}115 - 173.9T \ \text{J mol}^{-1}$

Adding these equations gives us, for the retort reaction,

$$ZnO(s) + C(s) = Zn(g) + CO(g) \tag{i}$$

$$\Delta G^\circ = 352{,}800 - 289.5T \text{ J mol}^{-1}$$

Using the expression

$$\Delta G^\circ = -RT \ln K$$

$$\log K = -18450/T + 15.12$$

where

$$K = p_{Zn} \cdot p_{CO}$$

In a process operated at normal atmospheric pressure, 101.335 kPa, the partial presssures of $Zn(g)$ and $CO(g)$ will both reach 50.67 kPa (0.5 atm) and $K = 0.25$ corresponding to a temperature of 900°C. Thus it is evident that continuous reduction of zinc oxide by carbon cannot take place below 900°C.

If we now calculate the equilibrium pressure of carbon dioxide for reactions (ii) and (iii), with the partial pressures of zinc and carbon monoxide each at 50.67 kPa (0.5 atm), and plot the values for the two reactions against temperature the curves shown in Fig. 12 are obtained [3].

At higher temperatures, due to the fact that reaction (iii) is much slower than reaction (ii), a steady state will be reached with the CO_2/CO ratio in the gas near to that determined by the zinc oxide equilibrium, and the driving force of the reaction will be the difference between this composition and that required for equilibrium in reaction (iii), or the distance between the curves in Fig. 12.

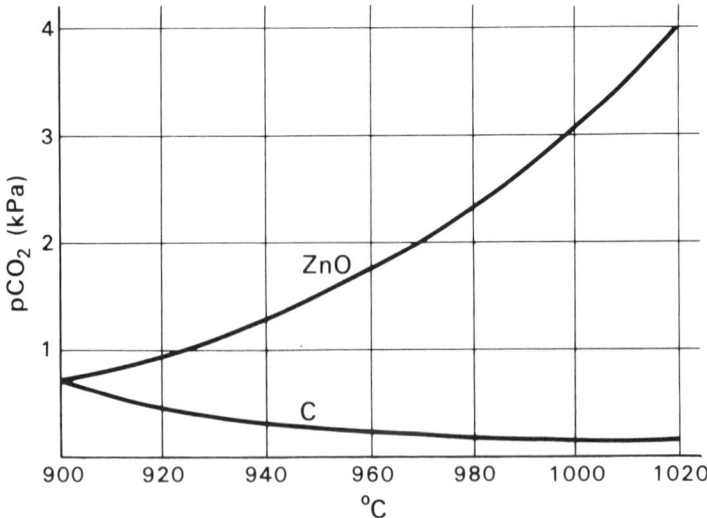

Fig. 12 – Equilibrium pressure of carbon dioxide (kPa) in the reactions $ZnO + CO$ (50.67 kPa) = Zn (50.67 kPa) + CO_2, $C + (CO_2) = 2CO$ (50.67 kPa).

The overall rate of reduction in the retort, however, is determined by the rate of heat transfer, and each element of charge reaches a temperature such that the rate of heat absorption due to the reaction is balanced by the gain in heat by thermal conduction.

For a number of reasons it is difficult to measure directly the temperature of the charge inside operating retorts, or the temperature and composition of the gas. From the calculations of the amount of lead volatilised and other indications, it would seem that in a horizontal retort using anthracite as reductant, the temperature of the charge over most of the distillation cycle lies between 980°C and 1000°C. The temperature of the vapours leaving the retort must also be in this range and since carbon is in excess in the charge, the ratio of carbon dioxide to carbon monoxide will be largely dictated by reaction (iii):

$$C(s) + CO_2(g) = 2CO(g)$$

and the content of carbon dioxide will be low. The gas consists of roughly equal proportions of zinc vapour and carbon monoxide, with approximately 1 per cent of carbon dioxide, and to cool it so as to recover the zinc as liquid demands special conditions. The dew point of such a mixture is 830°C and zinc will begin to condense as soon as this temperature is reached, but the operation is complicated by the fact that oxidation occurs as soon as the temperature begins to drop, since reaction (ii) then proceeds rapidly in the reverse direction. Unless, therefore, the rate of cooling is even greater, a large proportion of the condensing zinc will be oxidised, and excess 'blue powder', a dust consisting of particles of metallic zinc coated with zinc oxide, will be formed, which will not coalesce and form liquid metal. However, a limitation arises from the fact that if the temperature of the cooling surfaces drops below the melting point of zinc, solid zinc is formed again. Conditions must be maintained therefore to ensure that the temperature of the cooling surfaces lies above 419°C but that sufficient area is available to cool the vapours so rapidly that reoxidation is reduced to a minimum. Under such conditions the zinc condenses as liquid on the walls of the condenser and runs down to collect in the bottom, with less than 5 per cent reoxidised.

The size, shape, thermal conductivity and temperature of the condenser are therefore critical, and efficient condensation is possible only when these parameters are fixed within narrow limits.

Undoubtedly, the many problems which must be overcome to produce zinc thermally delayed the commercial production of zinc until much later than that of the other base metals.

4.2 THE HORIZONTAL RETORT PROCESS

The process first operated by Dony as described in Chapter 1 became the horizontal retort process which was used exclusively by the rest of the world and produced over 90 per cent of the total zinc until the electrolytic process was developed in 1917. Although it has played a major part in the growth of the zinc industry, it now appears that its days are numbered. In 1969, twenty-six plants were in operation producing about 15 per cent of the total available world capacity, but in 1982 the number had been reduced to five. The process is rapidly becoming obsolete, and is described fully elsewhere [4], but it exemplifies a number of the peculiar features of zinc distillation and merits a brief description.

The horizontal retort process as used in England

By 1850 the Belgian-type furnaces had been installed in the Swansea district, where suitable coal was plentiful, and for many years all the zinc smelted in this country was produced by the horizontal process. It was overtaken by more efficient methods and the last furnace was shut down in 1951. Figure 13 shows diagrammatically the last type of furnace used in this country at the Avonmouth plant of the Imperial Smelting Corporation.

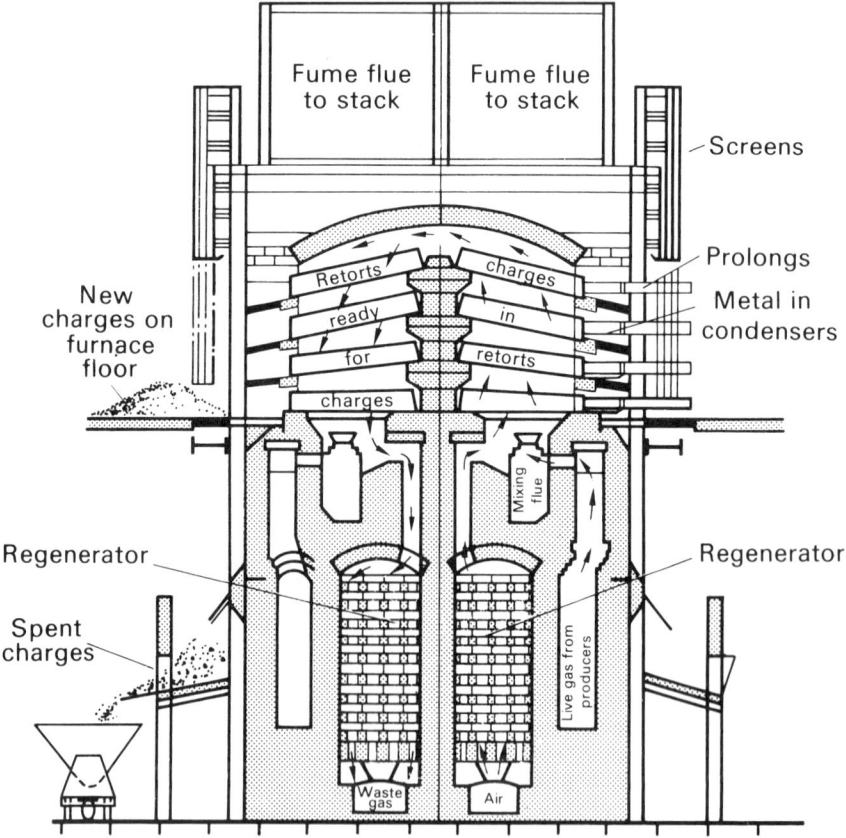

Fig. 13 — Section through horizontal retort zinc distillation furnace as used at the Avonmouth plant of the Imperial Smelting Corporation.

The retorts were arranged in the furnace in four rows on each side. Each section of a furnace contained 96 retorts, there being four sections to a furnace. The retorts were made of carefully selected clays which were blended, thoroughly mixed and conditioned. They were formed by extrusion under vacuum which de-aerated and improved the plasticity of the mix, and after drying were finally baked in special stoves, before being inserted at a temperature of 800°C into the furnace. The retorts were 1700 mm long and roughly elliptical in shape, the vertical axis being 270 mm and the horizontal 210 mm. The wall thickness was 32 mm and the capacity of a retort was 0.07 m³. Each produced on average 35 kg of metallic zinc per day.

The furnace was fired by hot producer gas fed into a flue running along one side of the furnace, where it was joined by air drawn up through the brickwork generator below, having been heated to 950–980°C. Combustion commenced in this flue and continued as the gases rose up past the retorts on one side of the furnace and down the other. The hot spent gases passed the regenerator below the second side to leave at a temperature of 570°C into waste-heat boilers. Every 20 minutes the passage of air and gas was reversed by a system of valves.

The furnaces were operated on a 24-hour cycle. The first operation at the beginning of the cycle was to replace any old retorts known to be leaking by hot new retorts. The condensers, also made of clay, were removed and placed on one side. The spent charge from the previous day's operations was then 'stirred out' by means of a scoop and rabbles being dropped into bins below the furnace. The new charge had been prepared and placed in front of the furnace the previous day. A variety of blendes was treated, preferably with a low iron content, since attack on the retorts was excessive if the iron content rose above 9 per cent. The blendes were roasted on down-draught sintering machines and the resulting sinter was crushed to particles of less than 6 mm to form a porous granular charge. To every 100 parts of sinter was added 30 parts of anthracite and 2–3 parts of salt, to assist condensation.

When all the residues from the previous day's production had been discharged, the new charge was moistened and then thrown manually into the retorts.

In view of the arduous nature of this operation attempts were made to adopt mechanical methods. Machines had been developed in Europe and in the United States which threw the charge into the retorts, generally off a moving belt. Although these were successful elsewhere, inferior distillation results were always obtained in this country, and the savings in labour could not be justified.

When the retorts were charged, the condensers were replaced on the mouths of the retorts and clayed in position. Over the condenser outlets were fitted the 'prolongs', which were cylindrical steel cannisters, 900 mm long and 130 mm in diameter, which acted as settling chambers for the vapours leaving the condensers, collecting most of the zinc escaping as zinc dust or 'blue powder'.

During the rest of the distillation cycle the temperature of the furnace was raised progressively so as to control the velocity of reduction at as constant a rate as possible. The final temperature reached at the end of the cycle was of the order of 1370°C and the furnace foreman showed a high degree of skill in choosing the correct temperature gradient, since if the charge was exhausted too early, excessive retort failure could occur.

The metal, which collected in the condensers, was removed at regular intervals, 7 to 9 kg being withdrawn at each tap. It contained considerable quantities of lead and iron, which increased as distillation proceeded. Cadmium was concentrated in the first metal to be withdrawn.

For sale the metal was required to satisfy the British GOB (Good Ordinary Brand) or American Prime Western specifications, which lay down the following limits:

	Prime Western (US)	GOB (UK)
lead	1.60%	1.35%
iron	0.08%	0.04%

The metal produced during the early stages of distillation satisfied these requirements, and was usually sold as produced, but the rest of the metal required refining by liquidation. This relied on the fact that the solubility of both lead and iron in molten zinc falls as the temperature is reduced. Liquation was usually carried out in a pool of metal held in a reverberatory furnace heated by either oil or gas. The temperature of the metal in the pool was maintained near to that of the melting point of zinc, and under such conditions excess lead and iron separated out and sank to the bottom of the bath. Overflow metal containing 1.1–1.2 per cent lead and 0.02–0.025 per cent iron could then be produced without difficulty, to satisfy the above specifications.

A considerable amount of skill was required to operate the furnaces satisfactorily. Owing to reasons which were never determined, furnace charges tended to work differently from day to day, and the temperature schedule used during the distillation cycle had to be varied correspondingly. With well-sintered blende giving a granular product containing less than 1 per cent sulphur, it was possible to eliminate 95 per cent of the zinc from the retorts. With good practice the zinc balance tended to be:

100 kg of zinc in charge	
metal	75 kg
blue powder	4 kg
dross	10–12 kg
zinc in residues	5 kg
losses – fume leakage, etc.	5 kg

Since the blue powder and dross from one cycle was added to the charge for the next, the recovery from new zinc fed to the retorts could reach 90 per cent, but it was frequently below this figure.

American practice with horizontal retorts

Until recent years natural gas could be obtained in the United States at a very low cost, and consequently the process was centred in Oklahoma and Texas where natural gas was so abundant that, at the prices then ruling, heat conservation by the use of air preheaters and waste-heat boilers could not be justified. As a result, the furnaces used in the United States were of simple construction without air regenerators. A 48-hour distillation cycle was used instead of that of 24 hours favoured in Europe, and although this reduced furnace output and increased fuel consumption, this was less financially serious than under European conditions. An advantage was that the furnace need not be driven so hard and this improved retort life and condensation efficiency.

The Overpelt process

A major modification to the horizontal furnaces was carried out at Overpelt in Belgium. Individual condensers and prolongs were discarded, and a common condensing chamber formed along each side of the furnace by suspending a removable sheet metal curtain running the length of the furnace and approximately 300 mm from the retort mouths. The vapours issuing from all the retorts on each side of the furnace were collected in the two collecting chambers thus formed. During distillation the heat loss from the sheet was sufficent to cool the vapours and condense most of the zinc, which was collected in a trough running the length of each side of the furnace, into a common sump from which the liquid metal could be removed. The carbon monoxide leaving the sump was scrubbed to remove any remaining zinc and then mixed with the gas fed to the firing flues of the furnace. A typical analysis of the scrubbed condenser gas would be:

CO	72–75%
CO_2	1–2%
N_2	4–6%
H_2	16–18%

with an energy value of approximately 10.5 MJ/m^3.

In orthodox furnaces this gas burns at the mouth of the condensers and is lost. Valuable use was made of it in the Overpelt system, and fuel consumption was reduced by 20 per cent.

The Overpelt system was adopted in Belgium, in the United States and in Mexico, but the additional complication it caused in furnace operation and its sensitivity to retort leakage were not offset by the saving in fuel consumption, and its use was not widespread.

Disadvantages of the horizontal retort process

A disadvantage of the horizontal retort process was the considerable amount of labour required, and many of the operations were hot and physically arduous.

Under the best United Kingdom conditions, to produce 1000 kg of zinc required 20 to 24 manhours — two to three times that needed in a modern electrolytic or blast furnace plant.

The energy required by the process was also high, in some cases almost twice that used by modern electrolytic and blast furnace plants (see Appendix 1).

The process suffers all the disadvantages of relatively small-scale batch operation, with high energy costs, and the labour required is exacting and demands conscientiousness of a high order. Although the process is still in use in Poland and in Mexico, all other furnaces have now been shut down. Since it was developed it has given good service, but it must now be regarded as obsolete.

4.3 THE VERTICAL RETORT PROCESS

The disadvantages of the horizontal retort process made its replacement inevitable. The first serious challenge came with the development of the electrolytic process in 1917, but at first many difficulties awaited solution, and several unsuccessful attempts were made to develop a new thermal method. In the late 1920s the New Jersey Zinc Company of Palmerton, Pa., in a programme of development work the intensity of which has probably never been equalled in the zinc industry, solved a series of formidable problems to develop a vertical retort unit, which operated continuously, and at high thermal efficiency.

The first major task was the construction of the retort itself, because the reduction of zinc oxide demands a considerable amount of heat. The basic reaction (i)

$$ZnO + C = Zn + CO$$

requires approximately 5.4 GJ per tonne of zinc produced, which is much greater than that necessary to reduce the oxides of lead, tin or copper. Assuming the vapours leave the retort at 1100°C, the additional amount of heat required to raise their temperature to this level is 1.2 GJ per tonne of zinc. In addition, heat has to be supplied to raise the temperature of the rest of the charge and to reduce other oxides present such as iron and lead. Thus, the total amount of heat to be passed through the retort wall to produce 1 tonne of zinc is of the order of 7.5 GJ, and it is essential that the heat conductivity of the walls of the retort should be high. Vertical retorts have been used in the coking industry since 1913, but for coal carbonisation the heat requirements are far lower, and silica brick could be used. The new material chosen was silicon carbide (carborundum) with a heat conductivity of 15.9 Wm^{-1} K^{-1} at 1300°C — nine times that of silica brick.

Retort design

Modern retorts, built of tongued and grooved carborundum brick 115 mm thick, are 1850 mm long and 305 mm wide internally. The heated side walls fit into glands in the end walls which alone are tied into the furnace setting. The method of construction is shown by the diagram of a horizontal section through one of the retorts in Fig. 14, and the general arrangement in Fig. 15.

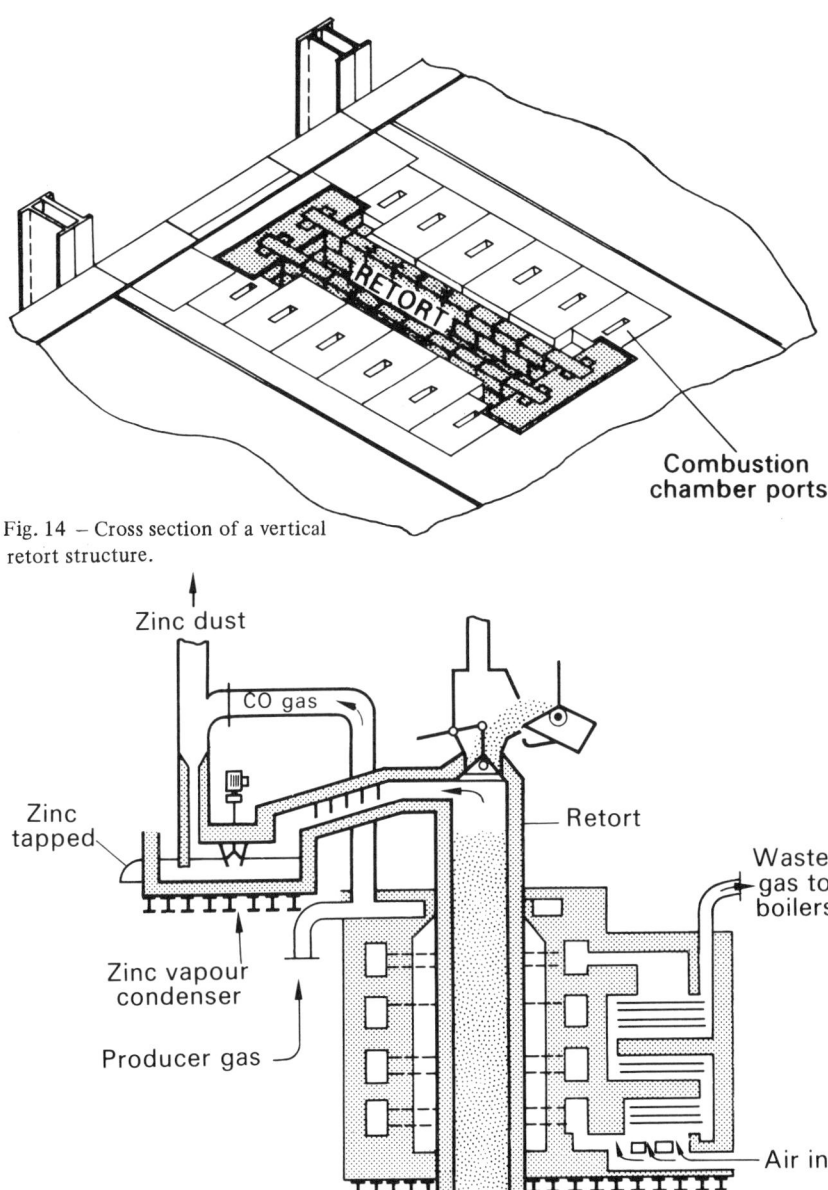

Fig. 14 – Cross section of a vertical
retort structure.

Fig. 15 – Diagram of a vertical retort plant.

The heated height of the tallest retorts yet built is 12 m, and such retorts can produce over 10 tonnes of zinc per day. As already stated, these 115 mm side walls are not tied to the setting except at the ends, and although heated to outside temperatures above 1300°C have a life of over three years. This is an excellent performance since, during the period, they are also subjected to internal abrasion owing to the passage of some 23,000 tonnes of charge.

Above the heated section of the retort is an unheated extension 3.65 m high, known as the 'eliminator', and the briquetted charge is fed into the top. The eliminator plays an important part in maintaining efficient condensation by permitting a certain amount of reversion of reaction (ii)

$$ZnO + CO = Zn + CO_2$$

This reduces the carbon dioxide content of the vapours. The zinc oxide formed is deposited on the briquettes to be carried down into the heated zone, to be reduced again. The eliminator also serves to reduce the lead content of the vapour leaving the retorts.

The retorts are built side by side in batteries of eight, each being fired and operated independently, and each can be rebuilt, if necessary, without disturbing the operation of its neighbours. They were heated by natural gas in the United States but by producer gas in Europe. The amount of heat required to produce 1000 kg of zinc varies somewhat with the type of concentrate treated, but is of the order of 20 GJ, of which approximately 30 per cent is supplied by the carbon monoxide-rich gas collected from the condensers. The mixture of gases is fed into burners at the top of the combustion chambers at each side of the retorts and drawn downwards. Heated air from the recuperators is introduced at various levels, and by careful adjustment uniform combustion can be assured; this is most important, and is the main factor in determining retort life. The furnace temperatures are rigidly controlled, and values as high as 1350°C have been run, but 1300°C is more usual. The waste gases leaving the bottom of the combustion chambers pass through counter current recuperators and heat the incoming air to 550°C before passing to the stack. The movement of material and gases is shown diagrammatically in Fig. 16.

Charge preparation and briquetting

The second major innovation of the New Jersey Zinc Company was a technique for producing a briquetted charge. The relatively large amount of heat required at a high temperature must be supplied to the charge by radiation, and to ensure this, a briquetted charge of a specific size is essential. Optimum conditions can be reached using a loaf-shaped briquette with rounded ends, 100 mm by 76 mm by 63 mm. Each such briquette can 'see' part of the hot wall and none is completely shielded from radiation — thus good conditions for heat transfer are reached.

Considerable care is taken in the manufacture of the briquettes since they

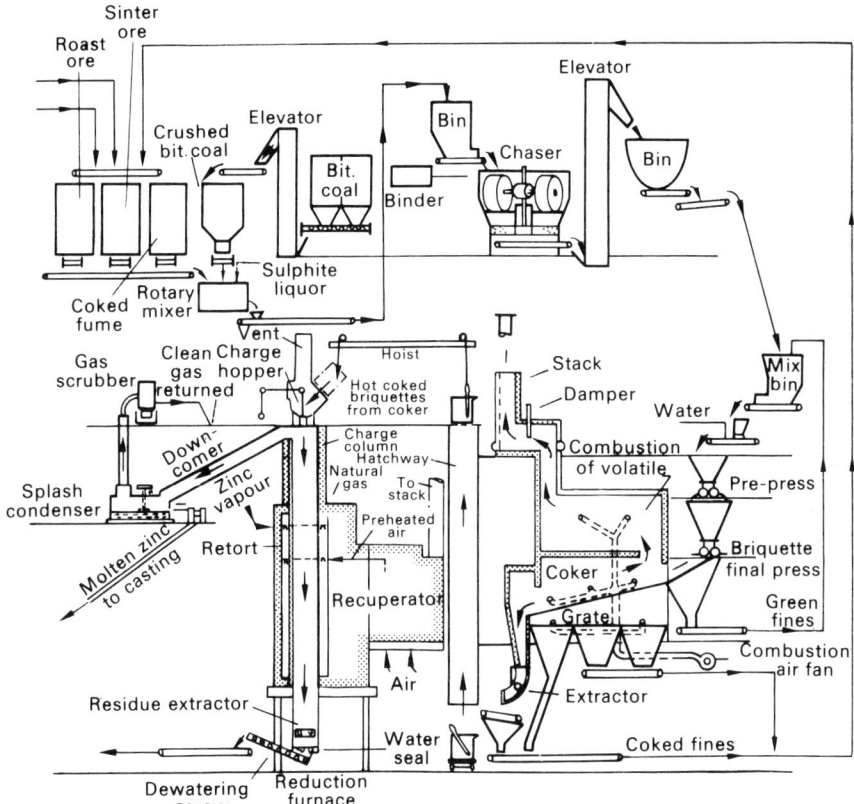

Fig. 16 – Flowsheet for a vertical retort plant.

have to stand up to exacting conditions. A mixture of sintered blende with a carefully selected bituminous coal of high agglutinating value, anthracite, and small proportions of ball clay and sulphite lye — a byproduct of the wood pulp industry — is thoroughly kneaded and plasticised in large mills and then formed in roll presses into loaf-shaped briquettes of the size required by the retorts.

Coking of the briquettes takes place in a furnace containing a series of downwardly sloping alloy steel step grates, and a slow reciprocating movement of alternate grates causes the briquettes to move down through the furnace. The temperature of the furnace is maintained by burning the volatile matter from the coal above the bed, and external heat is not required. After coking, the briquettes, which have lost most of the volatile matter in the coal and have a strong coke structure, are fed into buckets, which are weighed and then hoisted to the top of the retorts into which they are charged as required. Their temperature is of the order of 800°C.

Condensation

Special precautions have to be taken during condensation. As the back reaction

$$Zn + CO_2 = ZnO + CO$$

occurs rapidly on cooling, the vapours must be cooled as quickly as possible. To do this, the vapours leaving the eliminator pass through a short refractory lined downcomer into a rectangular chamber holding a bath of molten zinc, which is held at a controlled temperature – generally 500°C – by cooling coils, which can be raised or lowered, as required, into a well outside the condenser. At the outlet end of the condenser, a motor-driven carborundum shaft passes through the roof to drive a graphite impeller immersed in the zinc bath. This fills the chamber with a spray of zinc droplets which cool the vapours almost instantaneously to 500°C. The zinc vapour condenses as metal on the droplets, and as little opportunity for reoxidation is given, the conditions for condensation are almost ideal. Over 96 per cent of the zinc entering the condenser is produced as liquid metal, with blue powder production at less than 4 per cent.

After leaving the condenser, the gas is cleaned in venturi-type scrubbers and is fed to the combustion chambers of the retorts. Its volume is approximately 3 m^3 per minute per retort and since it contains 80 per cent carbon monoxide, it provides nearly 30 per cent of the heat required for distillation.

The analysis of the metal tapped from the condenser lies within the limits:

lead	0.1–0.2%
cadmium	0.03–0.06%
iron	0.005–0.015%

The future of the vertical retort process

The vertical retort process introduced a number of advantages over the older horizontal method but several inherent disadvantages still remained. Continuous distillation became possible for the first time, and thermal efficiency was improved, since most of the carbon monoxide produced by the main reduction reaction was returned to the retort combustion chambers. However, capital costs per unit of zinc produced are high, as are maintenance costs. The process is dependent on the quality of the bituminous coal used to give the coke structure of the briquettes, and highly agglutinating coals of the type required are not widely available. Probably the most serious disadvantage is that the process is restricted to low iron concentrates, because, if the iron content of the sinter fed to the retort rises above 10 per cent, plates of iron begin to form in the retorts, seriously interfering with the flow and discharge of the briquettes. The spent briquettes are bulky and present a disposal problem. A detailed description of recent vertical retort practice is given in the literature [5,6]. Two plants were built in the USA and one each in the UK, France, Germany and Japan, but two of the original installations have now been shut down and it seems unlikely that others will be built in the future.

4.4 THE ST. JOSEPH ELECTROTHERMIC PROCESS

In 1926, at the same time as the New Jersey Zinc Company was developing the vertical retort process, the St. Joseph Lead Company also began work on a continuous method of zinc distillation, but with a different approach. In both the horizontal and vertical retort processes the heat applied externally has to pass through a retort wall, placing an exacting demand on the refractories. To try to avoid this, attempts had been made to supply heat to the charge internally, the most successful being developed by De Laval at Trollhatten, Sweden, in 1898, in which the charge was heated by an electric arc. Although some furnaces using this principle were built, a satisfactory method of condensation was not developed and there were difficulties with slagging of the spent charge.

The principle adopted by the St. Joseph Company was to develop the heat required by using the electrical resistance of a vertical column of the charge, power being supplied through carbon electrodes [7]. The method showed promise, many engineering difficulties were overcome, and furnaces capable of continuous operation were built. A charge of graded sinter and coke in approximately equal proportions by volume was used, but at this time no attempt was made to condense the zinc vapour leaving the furnace shaft, and it was allowed to burn in an adjoining chamber producing pigmentary zinc oxide.

These furnaces established the practicability of resistance heating, and attention was directed to the problem of condensation. After a number of trials a method was developed of drawing the vapours from the shaft through a pool of molten zinc. This was a brilliant innovation, and was the first application of molten metal as a cooling and condensing medium. It was later used in the lead splash condenser of the Imperial Smelting furnace, and the zinc splash condenser in vertical retort operation. For the first time, large quantities of zinc vapour could be condensed to liquid metal, with the minimum production of blue powder.

The first commercial condenser operating on the new principle was fitted to an oxide furnace and showed considerable promise. It had a daily capacity of 5 tonnes of zinc, and today units producing over 80 tonnes of zinc per day using a single condenser are in operation. The broad design of the latest St. Joseph furnaces operating at the Josephtown smelter of the St. Joseph Minerals Corporation [8] is shown in Fig. 17.

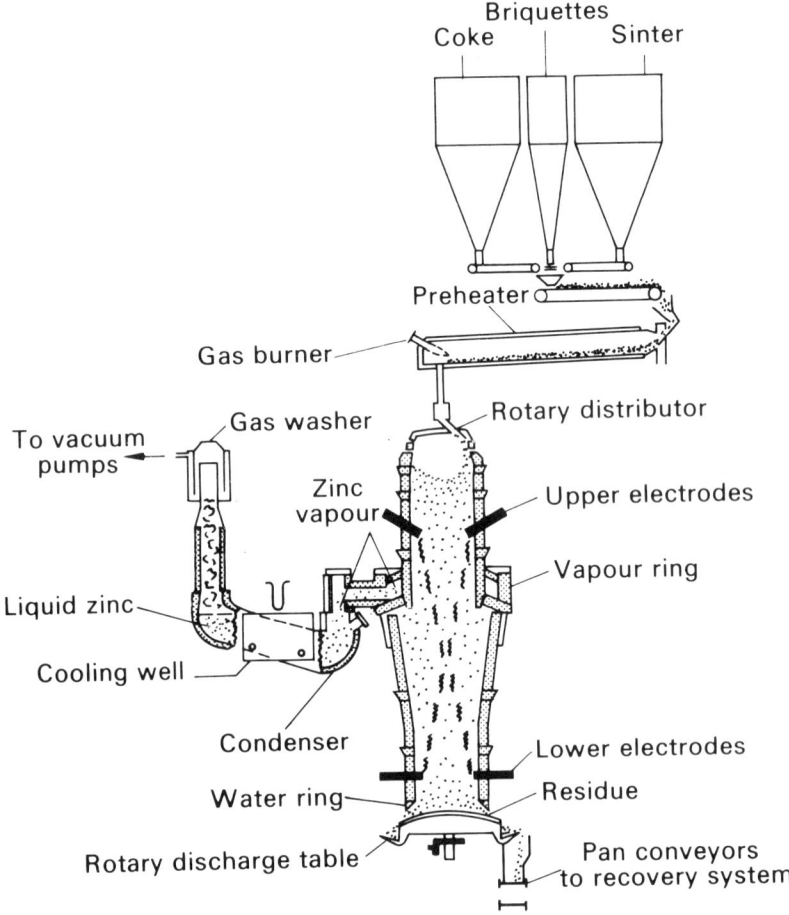

Fig. 17 – Diagram of an electrothermic zinc furnace used by the St. Joseph
Lead Co.

The furnace is built of a number of firebrick sections each individually supported. The vapours produced in the furnace collect in an annular ring from which they are drawn off into the condenser at a temperature of 800°C–850°C. Silicon carbide bricks are used to line the ring and the condenser. The electrical current heating the charge is passed through the furnace via eight pairs of electrodes protruding into the furnace at top and bottom. The charge of coke and sintered blende, heated by carbon monoxide gas to 750°C, is fed into the top through a rotating distributor, and discharged from the bottom at a controlled rate by means of a rotating table, assisted by mechanically operated water-cooled rakes which are periodically pushed into and withdrawn from the furnace. The operation of both the table and the rakes is governed automatically by signals from a gamma ray charge level indicator at the furnace top.

Approximately 1250 kW fed through each pair of electrodes give a maximum furnace power consumption of 10,000 kW. The voltage between the top and bottom electrodes varies from 200 to 250 volts, depending on the conductivity of the charge. The power consumption of the furnace and its auxiliaries is of the order of 3050 kWh per tonne of metal produced. In addition, approximately 0.6 tonne of coke is required.

The condenser consists of two vertical legs connected by a sloping section inclined at 22° to the horizontal; the internal area of this section is 2 m². It is connected to an external cooling well of zinc from which heat is abstracted at a controlled rate by varying mechanically the immersion of hairpin coolers carrying water, which maintain the temperature of the bath at 480°–500°C. Approximately 93 per cent of the zinc entering the condenser is condensed to liquid metal.

From the top of the exhaust leg the vapours leaving the condensers are scrubbed by high velocity water sprays and any remaining zinc removed. The clean gas, with an approximate composition of carbon monoxide 79 per cent, carbon dioxide 3 per cent and nitrogen 18 per cent, then passes through vacuum pumps operating at a suction of 33.3–40.0 kPa (250–300 mm) of mercury, providing the motive power to draw the vapour from the furnace through the pool of liquid zinc in the condenser and the scrubbing system. The production of carbon monoxide averages 1.42 mol per mol of zinc vaporised, or 1.55 mol per mol of zinc produced. The gas, with a heating value of 9.3 MJ/m³, is a valuable source of energy, 60 per cent being used for charge heating and the remainder for power generation.

The residues from the furnace are subjected to extensive treatment, which raises the overall zinc recovery to 95–96 per cent. Magnetic separation removes high iron material low in zinc, and then 90 per cent of the free coke is recovered on pneumatic tables, in a form which can be returned to the furnace. Finally heavy media separation is used to recover a zinc fraction which is fed back to the sintering plant.

Despite the considerable technical and engineering skill shown in the development of the St. Joseph process, its application has been limited, although other electrothermal units have been built in the USSR and in Japan, and a small unit in Germany. The total energy requirement of the process is high, and although units producing over 100 tonnes of metallic zinc per day are in operation, this is considerably less than the output of most zinc–lead blast furnaces. The capital cost of electrothermal units per unit of zinc produced is relatively high. The iron content of the charge must be limited and therefore the process demands high grade concentrates. In consequence it seems unlikely that the process will play a major role in the future development of the zinc industry.

REFERENCES

[1] Maier, C. G., Zinc smelting from a chemical and thermodynamic viewpoint, Bulletin No. 32, US Bureau of Mines, 1930.

[2] Barin, I. and Knacke, O., *Thermochemical Properties of Inorganic Substances,* Springer-Verlag, 1973.

[3] Lumsden, J., The physical chemistry of the zinc blast furnace, *Proceedings of the Metallurgical Chemistry Symposium, Brunel University and the National Physical Laboratory July 1971,* Paper 44, pp. 533–48, HMSO, London, 1972.

[4] Foster, M. K., Horizontal retort and acid plant, Asarco, Mexicana, *World Symposium on Mining and Metallurgy of Lead and Zinc,* American Institute of Mining and Metallurgical Engineers, Vol. 2, p. 463, 1970.

[5] Haldacre, G. F. and Peirce, W. M., The vertical retort process, *Zinc,* ed. C. H. Mathewson, p. 252, Reinhold, New York, 1959.

[6] Fetterolf, L. D., Bechdolt, W. R., Stilo, V. and Motto, J. A., The vertical retort zinc smelter at Depue, *World Symposium on Mining and Metallurgy of Lead and Zinc,* American Institute of Mining and Metallurgical Engineers, Vol. 2, p. 512, 1970.

[7] Weaton, G. F., The electrothermic process of the St. Joseph Lead Company, *Zinc,* ed. C. H. Mathewson, p. 270, Reinhold, New York, 1959.

[8] Lund, R. E., Winters, J. F., Hoffacker, B. E., Fusco, T. M. and Warnes, D. E., Josephtown electrothermic zinc smelter, *World Symposium on Mining and Metallurgy of Lead and Zinc,* American Institute of Mining and Metallurgical Engineers, Vol. 2, p. 549, 1970.

5

The zinc blast furnace

Whilst the vertical retort and the electrothermal processes were being successfully developed in the United States to overcome a number of disadvantages of the long-established horizontal retort method, a more revolutionary approach was being made by the Imperial Smelting Corporation at Avonmouth in England.

5.1 HISTORICAL BACKGROUND

Many unsuccessful attempts had been made in the past to develop a blast furnace process for zinc production, but all had failed from an inability to overcome the problem of condensation. Gases leaving a horizontal or vertical retort, containing, approximately, zinc vapour 45 per cent, carbon monoxide 50 per cent and carbon dioxide 1 per cent, were difficult enough to condense without excessive reoxidation. From a blast furnace the gas could be expected to contain only 6–7 per cent zinc and 12 per cent carbon dioxide and the problem was thus even more forbidding.

Prior to 1939 a study was undertaken at Avonmouth by the then National Smelting Company, using a small furnace which was fed with sintered zinc oxide and coke. Air was blown in at both top and bottom of the furnace and the vapours were withdrawn from a point near the centre. Shock-chilling of the vapours in a water-cooled tube condenser showed that a zinc dust could be formed with a high metallic zinc content, and thus at rapid rates of cooling the actual proportion of zinc reoxidised by the back reaction was not great. It was sufficient, however, to prevent coalescence of the dust when subsequently heated, and as liquid zinc could not be made, a radically different method of shock-chilling the vapours was required.

The problem was solved by a proposal originated by L. J. Derham [1] who suggested that a spray of molten lead droplets at a temperature slightly above the melting point of zinc should be used to scrub the vapours immediately they entered the condenser. Under these conditions the vapours would be cooled before extensive reoxidation could occur to give liquid, not solid, zinc which would dissolve in the lead.

By withdrawing the lead continuously from the condenser and cooling it, the saturation point of zinc in lead could be reached and the molten zinc, having separated out above the lead, could be removed. The cooled lead was then to be returned to the condenser to form again the shock-chilling spray and to take up more zinc. The principle involved can be appreciated by a study of the lead-rich end of the Zn—Pb phase diagram (Fig. 18). The viability of this technique was tested in practice and promising results obtained, although the commencement of hostilities in 1939 halted further work on the new process.

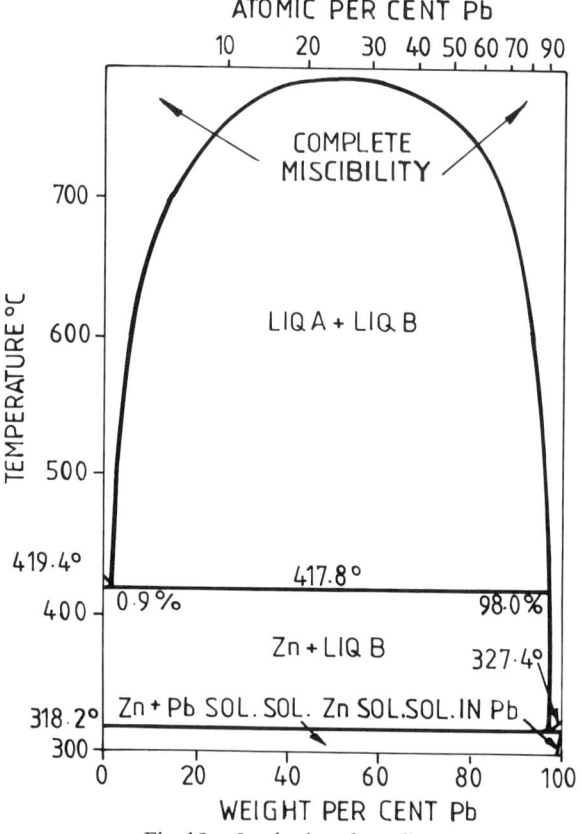

Fig. 18 – Lead–zinc phase diagram.

In 1947 an intensive attack on the problem was launched. Thermodynamic calculations indicated that it should be possible to reduce the zinc oxide content of the run-off slag to low levels without producing metallic iron, which was important since a process requiring simultaneous reduction of the iron oxide in the charge would consume carbon needlessly. This was confirmed in practice, and it was also shown, as theory predicts, that lead oxide is reduced in the upper levels of the furnace by the carbon monoxide present. Since this reaction is exothermic, additional carbon is not required and the capacity of the furnace to produce zinc is largely unaffected. The molten lead so produced ran down through the charge, acting as a collector for copper and precious metal. Thus, the furnace could be used to treat simultaneously both lead and zinc concentrates, and its viability was increased.

Although the potential of the method was soon demonstrated, a long period of intense development was required before the process was capable of smooth continuous operation. One of the most serious problems was caused by heavy deposition of zinc oxide in the top of the furnace and the flue leading into the condenser. The gas leaves the top of the charge under conditions in which the reaction

$$ZnO(s) + CO(g) = Zn(g) + CO_2(g)$$

is at equilibrium, and, as expected, as soon as its temperature begins to drop, reversion takes place with deposition of zinc oxide on all surfaces exposed to the gas. This builds up rapidly and the furnace becomes inoperable.

Eventually a simple solution was evolved, deduced from the physical chemistry of the system. A typical furnace gas at equilibrium can be shown to have the characteristic that, if it reacts adiabatically with carbon, despite the fact that its carbon dioxide content is lowered, its temperature will also fall, so that the gas is more prone to deposit zinc oxide; conversely, however, if it reacts with oxygen the rise in temperature more than compensates for the increase in carbon dioxide and the gas can now reduce zinc oxide. Thus, the remedy was suggested by S. E. Woods [2] that to prevent zinc oxide deposition, further air should be added to partially combust the gas. The principle proved successful and, as a result, the furnace design was greatly simplified. Although, with the addition of 'top air', some deposition of zinc oxide still occurs, its growth is largely restricted to the inlet to the condenser, from which it can be removed.

Further modifications were found necessary. At first, down-draught machines, built to prepare sinter for the Company's horizontal and vertical retort operations, were used to prepare the lump sinter required for the blast furnace, but these proved inadequate for their new duty. There was growing pressure to increase the lead content of the charge since this improved the economics of operation, though it clogged the grate bars. A decision was taken to reverse the flow of air through the pallets on the machines and to blow upwards. This had already proved possible on the galena sintering machines at the lead smelter at Port Pirie in Australia, and eventually good quality sinter was made by up-draughting from blende mixtures containing considerable quantities of galena and other lead compounds. The gas leaving the machines contained 6–7 per cent sulphur dioxide which was converted to sulphuric acid without difficulty.

Confidence in the process having been established, a unit at the Swansea Vale works of the National Smelting Company was completed in 1960 with a shaft area of 17.1 m^2 to produce about 33,000 tonnes of zinc and 15,000 tonnes of lead annually — figures which were later greatly exceeded. A number of refinements were introduced — air preheat temperatures were raised to 750°C, more extensive automatic control was applied and the efficiency of charge proportioning was increased. The furnace came on line in 1960 and almost immediately a second furnace was commissioned for the Sulphide Corporation at Cockle Creek in Australia, in which it was decided to use one condenser instead of the two provided for the Swansea plant. This was a further important simplification. A description of the new process was first published in 1956 [3], and full details of established plants have been published in recent years [4].

5.2 ZINC BLAST FURNACE DESIGN

The general layout is shown in Fig. 19. The feed at the top consists of solid sinter and heated coke (at some plants briquettes of dross and other secondary materials are also charged). Air, preheated to 950°C–1050°C, is blown in at the bottom. The hot gas leaving the top carries zinc vapour and some volatilised sulphides; liquids tapped from the bottom consist of slag and lead bullion. Small quantities of matte or speiss occasionally occur. Zinc is scrubbed from the emerging vapour by liquid lead in the condenser, and liquid zinc product separates from lead in the cooling circuit. Main aspects of the smelter are discussed in the following paragraphs.

Furnace data

The design of furnace is based essentially on that of the lead blast furnace, but the different operating requirements impose a number of major modifications. A rectangular shape with curved ends for the horizontal cross-section is adopted, since the distance between opposite tuyeres must be limited to allow blast penetration across the whole hearth. The bottom section of the furnace is water-cooled, in view of the high temperatures attained in the combustion and slag melting zones. Originally, this section was formed of two tiers of water jackets but, as these tended to distort and leak, most furnaces now use a monolithic casing constructed of steel plate cooled externally by a falling film of water. The tuyeres project through the casing into the furnace, and due to the high air preheats now used, they are subject to severe conditions and their design is critical. The temperature in the raceway in front of each tuyere can exceed 1600°C, and to safeguard the tuyere nozzles, it is essential that cooling is sufficient to cope at all times with a very high heat flux.

The steel tuyeres are annular in construction, forming concentric water-cooled tubes, and the cooling water is forced down the inner annulus to the tuyere nozzle, from which it flows back to an exit outside the furnace. The velocity of the water is increased by imposing a spiral flow pattern so that steam generation cannot take place and maximum heat transfer is obtained.

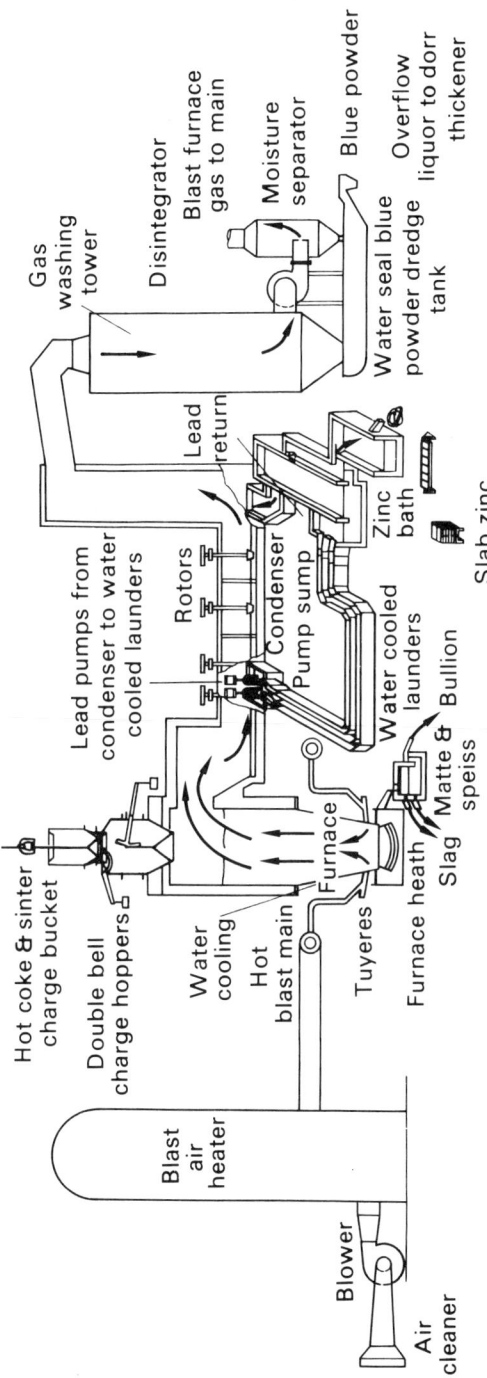

Fig. 19 – General layout of zinc–lead blast furnace plant.

After successful development work the general design of the furnace was established in the early 1960s. This design of a typical furnace, with 26 tuyeres, is shown in Fig. 20.

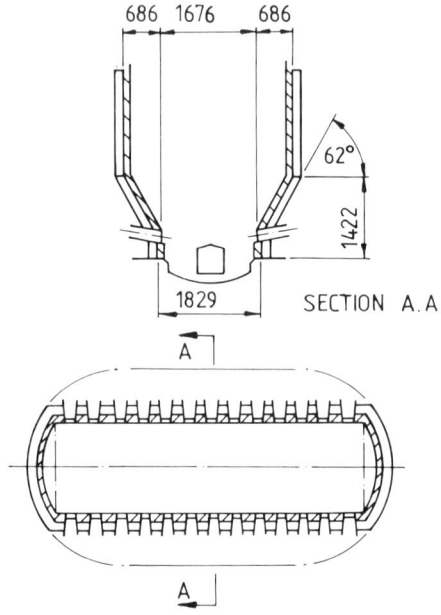

Fig. 20 – The 1963 furnace – 17.2 m² (modified from Harris [5]).

Furnaces of this type originally had a production capacity of 33,000 tonnes of zinc per annum. The present production capacity of a furnace with the same upper shaft area is approximately 80,000 tonnes of zinc per annum. The greater part of this improved output has been due to increased air blowing rate, which in practice equates with production rate. The improvement has required modification to the hearth region so that on a typical furnace the width at tuyere level inside the casing is now 2550 mm, the distance between tuyere noses is 1940 mm, the tuyere spacing is 710 mm, and there are 16 tuyeres with a bore of 14 mm. The blast air, heated on several modern furnaces to at least 950°C, creates raceways in front of the tuyeres, similar but smaller in size to those in iron blast furnaces, and each operator aims to get the best pattern of distribution of the raceways, and the intense combustion zone surrounding them, so that the whole of the furnace bottom works as evenly as possible.

The raceway acts as a distributor for blast air, and for combustion of coke; it also promotes heat and mass transfer between itself and the slag pool. Gammon [6] examines the factors influencing the development of raceways, and their influence in promoting heat and mass transfer.

5.3 SHAFT REACTIONS

In an analysis, first developed by Lumsden [7], of the reactions which must take place in the furnace shaft, a number of zones can be distinguished although these are not necessarily sharply defined. They are represented diagrammatically as Fig. 21.

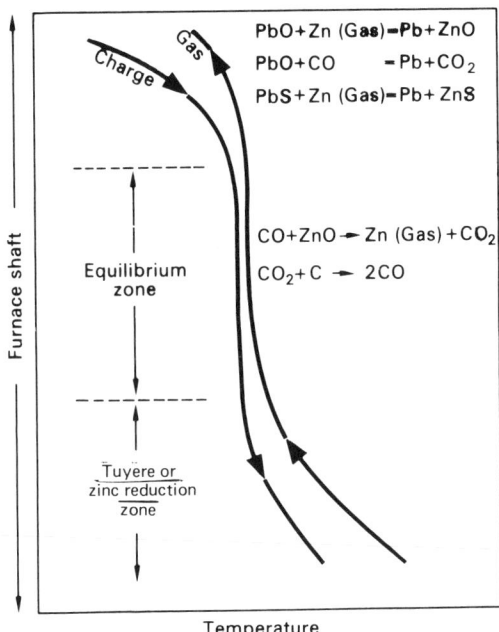

Fig. 21 – Approximate division of blast furnace into zones.

As a starting point, one can assume that at some level in the furnace neither reduction of zinc oxide nor oxidation of zinc vapour takes place, and at this level the composition of the gas corresponds to equilibrium at the temperature of solid zinc oxide charge. This temperature is of the order of 1050°C under the the conditions ruling, and above this level two zones can be distinguished.

Zone 1: Charge heating zone

The charge consists of a mixture of hot coke (temperature approximately 800°C) and cold or warm sinter, and its mean temperature on entering the furnace is therefore of the order of 400°C. Its temperature is raised by direct transfer of heat from the gas and also by some reoxidation of zinc vapour. It also receives some heat from the combustion of gas by the air added above the furnace charge.

Zone 2; Reoxidation zone

When the solid charge temperature is close to the gas temperature there is a possibility of two reactions occurring in thermal balance: the reduction of carbon dioxide by carbon, absorbing heat; reoxidation of zinc vapour, liberating heat. Thus, there must exist an extended 'equilibrium zone' in which the conditions, while not far from equilibrium, continue to oxidise zinc. This zone can be considered a thermal reserve zone; whereas the interchanges occurring in the zones both above and below it are such that only a limited length can be allocated to them, the equilibrium zone, with its relatively slow changes in gas composition, can adjust itself to the difference in reaction height between the total provided by the furnace design and the length required for the preceding and succeeding stages.

A number of relatively reducible species can be reduced in the zones before the reduction of zinc oxide. Lead oxide is unstable virtually throughout the whole of the zones; lead sulphate and calcium sulphate are reduced to sulphide.

Zone 3: Solid reduction
There is a substantial range of temperatures between that at which solid zinc oxide is in equilibrium with the furnace gas richest in zinc vapour content, and the lowest temperature at which liquid slag is obtained. This range is increased by choosing a slag with a high melting point. The zone in which solid temperature lies within this range is one in which reduction of zinc takes place. It is desirable in the interests of zinc elimination that the proportion of the zinc oxide input reduced in the solid state should be as great as possible; this will depend on how much hotter the gas stream can be, compared with the solid, before slag melting becomes significant, and on the melting point of the slag.

In the solid reduction zone, the gas will contain a considerable proportion of lead and lead sulphide vapour. Circulation of these species must occur with condensation on colder charge and return to the hotter zones below. This circulation provides an additional mode of heat transport.

Zone 4: Melting, reduction and combustion
In this zone the main task is the reduction of the remaining zinc oxide, and the melting of the slag. The zinc oxide entering this zone is presumably dissolved into, and then reduced out of, the slag. This is not necessarily a slow process, but the amount which occurs depends largely on the time available before the slag reaches the relatively quiescent pool below the tuyeres. In the raceways formed at the tuyeres there is circulation of both solid and liquid phases, which tends to raise liquid slag temperatures and increase residence time. In addition, in the raceway and its near vicinity, the consumption of oxygen takes place developing the heat necessary to operate the furnace.

Slag composition
From the furnace model so outlined it follows that little reduction of zinc oxide occurs until the charge has left the equilibrium zone relatively low down in the furnace. As the charge leaves this zone it moves into conditions increasingly

favourable for reduction and the main duty of the furnace – that of zinc oxide reduction – occurs. As the sinter continues to descend through this stage its temperature rises until the liquidus of the sinter gangue is reached and melting begins. The liquid then runs down into the slag pool below the tuyeres, where opportunity for further chemical reaction or heating is diminished. Thus, the greater the gap between the temperature at which the sinter leaves the equilibrium zone – approximately 1050°C – and its melting point, the more complete will be elimination of zinc. The analysis of the slag produced varies with the feed, but in general, slag composition lies within the following range (per cent):

FeO ... 30–42
SiO_2 ... 16–21
Al_2O_3 ... 5–10
ZnO ... 5–10
PbO ... 0.6–1
S ... 1–3

In the operational life of the zinc blast furnace it is known that to obtain good zinc elimination with a good fuel efficiency it is helpful to incorporate sufficient lime in the charge to produce a slag with a fairly high CaO/SiO_2 ratio, say of the order of 1.5. More recently it has been observed that with slags of lower CaO/SiO_2 ratio the zinc elimination is improved when the slag pool is kept in better contact with the furnace gases, formed by combustion of tuyere air, by increasing the retention time of the slag and maintaining as high a slag level as possible at the tuyere region. One advantage of keeping the CaO/SiO_2 ratio low is that it leads to a smoother operation. High CaO/SiO_2 ratios, although assisting good zinc elimination, are thought to be associated with operational difficulties attributable in part to above-normal accretion formation, and possibly also, to the formation of some metallic iron. Certainly where over-reduction has occurred it is known that difficulties arise in discharging slag, and it is obviously beneficial to know the limitations of the process with regards to the problems of iron formation.

The two main metal reduction reactions in the blast furnace are

$$ZnO(s) + CO(g) = Zn(g) + CO_2(g) \qquad (1)$$

and

$$PbO(s) + CO(g) = Pb(l) + CO_2(g) \qquad (2)$$

Copper is also reduced and the standard free energy diagram (Fig. 11, Chapter 4) is a reminder of the thermodynamics of the reduction processes. It emphasizes the point that zinc is evolved in the gaseous state when zinc oxide is reduced by carbon and carbon monoxide at normal pressures.

Both reactions (1) and (2) are fast. The second reaches completion in the upper shaft of the furnace. However the first, being very temperature-dependent and requiring a large amount of heat, is not completed until the charge reaches the melting zone in the lower shaft of the furnace. In the active tuyere zone the gangue content of the charge becomes molten, and a substantial portion of zinc is reduced from zinc oxide dissolved in liquid slag at a temperature close to that of the slag liquidus. The slag discharged from the furnace contains at least the equilibrium quantity of zinc oxide dictated by the reaction

$$ZnO \text{ (in slag)} + CO(g) = Zn(g) + CO_2(g) \tag{3}$$

The equilibrium constant is given by

$$K_3 = \frac{p_{Zn}}{a_{ZnO}} \cdot \frac{p_{CO_2}}{p_{CO}}$$

where p_{Zn}, p_{CO_2} and p_{CO} are the partial pressures of the components and a_{ZnO} is the activity of zinc oxide dissolved in slag. If iron is not to be reduced, the CO_2/CO ratio of the gas in the slag reduction zone must be such that it does not reduce iron by the reaction

$$FeO \text{ (in slag)} + CO(g) = Fe(s) + CO_2(g) \tag{4}$$

Therefore the maximum extent to which zinc can be reduced from the slag will be governed by the equilibrium in the reaction

$$ZnO \text{ (in slag)} + Fe(s) = FeO \text{ (in slag)} + Zn(g) \tag{5}$$

In the following equation $K_5 = p_{Zn} (a_{FeO}/a_{ZnO})$ (with γ representing activity coefficients)

$$Zn \text{ (in slag, mol\%)} = \frac{p_{Zn}}{K_5} \frac{FeO \text{ (in slag, mol\%)}}{(\gamma_{ZnO}/\gamma_{FeO})} \tag{6}$$

Using the data of Barin and Knacke [8] the equilibrium constant of reaction (5) with ZnO and FeO (wüstite) referred to the solid is calculated as

$$1400 \text{ K} \quad K_5 = 0.421$$
$$1500 \text{ K} \quad K_5 = 1.31$$
$$1600 \text{ K} \quad K_5 = 3.52$$

Taking 1500 K as a typical melting point of a blast furnace slag and assuming a slag close to wüstite separation with $a_{FeO} = 1$, and $p_{Zn} = 0.1$

$$a_{ZnO} = 0.1/1.31 = 0.076$$

Activity determinations for zinc oxide and ferrous oxide dissolved in liquid silicate slags have been carried out by several workers independently. Although covering a wide range of $FeO-CaO-SiO_2-ZnO$ compositions the results of these authors are in reasonable agreement [9]. A typical activity coefficient of zinc oxide dissolved in slag, referred to the solid, is about 2.5 so that $N_{ZnO} = 0.076/2.5 = 0.03$; in other words it should be possible to remove zinc from liquid slags down to a concentration of about 3 mol% before there is a tendency to reduce iron from the liquid slag. Lumsden has calculated the data given in Fig. 22 which relates to the typical situation in a zinc blast furnace.

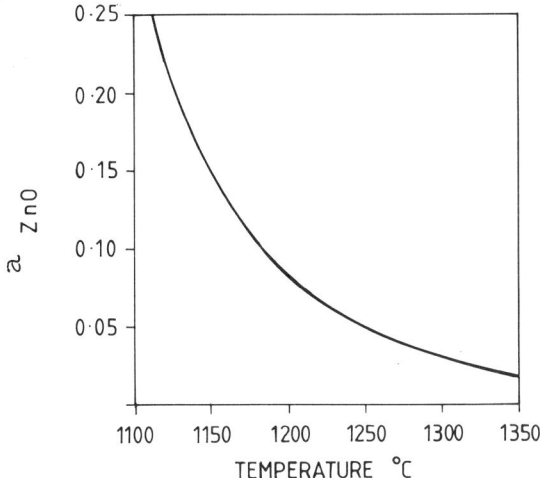

Fig. 22 – Limit of zinc oxide activity in blast furnace practice to avoid iron reduction.

The theoretical and experimental studies on zinc oxide activity in molten slags support the early operational findings that high CaO/SiO_2 ratios can be beneficial in lowering the zinc content of the slag. However, in the introduction to this section it was noted that operating blast furnaces do not now follow this high temperature slag approach. A more economic policy has been to operate at a lower basicity with slag melting points at about 1200°C in place of the 1250–1300°C melting point slags experienced at higher basicities. Under these conditions, although theory predicts a higher equilibrium zinc content of slag, in practice, owing to a lower slag fall and improved gas–slag contact in a more active tuyere zone, a higher overall recovery of zinc is now obtained.

In a recently presented paper by Harris et al. [10], improvements in recovery from the adoption of this technique were presented. Typical furnace results indicate that although there has been little change in the zinc assay of reject slags, the distribution of zinc in slag as a fraction of new zinc charged to the furnace has dropped from about 7.5 per cent to 4.5 per cent.

Further development plans and operational improvements to increase zinc recovery, whilst avoiding iron reduction, are also discussed and it is confidently expected that progress will continue in the next few years.

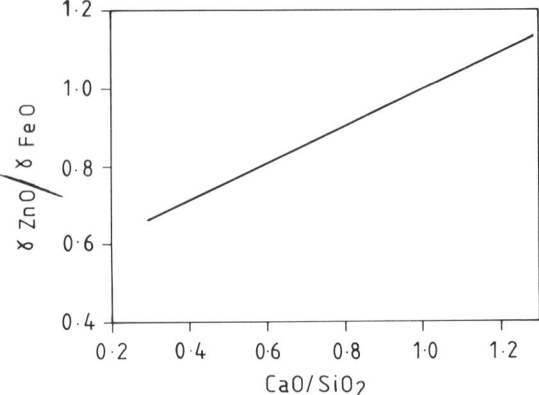

Fig. 23 – Variation of $\gamma_{ZnO}/\gamma_{FeO}$ with CaO/SiO_2 for typical blast furnace slags.

Coke requirements

The functions of coke in the furnace are to maintain the permeability of the charge, to provide the energy to operate the process and melt the gangue, and to reduce zinc oxide. Physically it should be sufficiently strong to resist abrasion in its passage through the coke preheaters and the furnace. A low volatile matter ($<$ 2 per cent) is desirable, since the hydrogen content of volatiles causes condensation difficulties, but the most important criterion is reactivity, low reactivity being preferred.

The reactivity of a coke is a measure of the rate at which carbon dioxide is reduced at a given temperature by the reaction

$$CO_2(g) + C(s) = 2CO(g)$$

The reactivity of metallurgical cokes varies considerably and it is a factor of great importance to the zinc blast furnace. The variations arise from a number of causes which are dependent to some extent on the coal used, but coking conditions have a considerable influence. High coking temperatures and length of time in the ovens favour the production of low reactivity coke.

With a coke of high reactivity the ratio of carbon monoxide to carbon dioxide in the furnace gas rises, and top temperatures fall, as a result of their loss of sensible heat due to endothermic reduction of carbon. The reduction zone tends to move upwards, which cools the hearth zone and can cause slagging difficulties. The combined effect of less carbon reaching the tuyere zone and a lower slag temperature increases the zinc loss in the slag. To compensate for these effects an increase in carbon–zinc ratio must be made, raising the fuel cost per unit of zinc. Metallurgical coke is a relatively scarce commodity and blast furnace superintendents have frequently little choice in the material they buy, but the advantages of low reactivity coke are appreciable.

A number of methods of producing formed coke, in which coal is briquetted before coking, are under examination. Since these can use a wide variety of coals and are not restricted to the narrow range from which conventional metallurgical coke is made, the successful development of such processes could widen the application of the zinc–lead blast furnace. The formed coke produced so far tends to be of higher reactivity than conventional oven coke, particularly from those processes which use pitch as a binder, and techniques for reducing this are required. Chemical inhibitors such as boric acid have some effect and their use is possible.

Before being fed to the furnace, coke is preheated to approximately 800°C in coke preheaters, which are vertical shafts down which the coke is fed against a rising stream of combustion gases produced by burning scrubbed gas from the condenser in an external chamber. Combustion conditions are automatically controlled to give low oxygen levels in the gases, otherwise some combustion of coke occurs in the preheater. The hot coke leaving the bottom is fed by charge buckets to the furnace bells. Its sensible heat improves the thermal efficiency of the furnace.

Coke consumption
Most furnaces are blown at a fixed rate determined by the capacity of the blower and the gas-handling equipment; thus the amount of carbon burnt per day is constant. The amount of zinc which can be produced from a furnace with a given blowing rate and, therefore, carbon consumption depends on the amount of slag which has to be made, which is determined by the grade of charge treated and the ash content of coke. In the early days, when air preheat rarely exceeded 700°C, experience showed that the heat required to melt one tonne of slag was 0.232 times that for reducing zinc oxide to produce one tonne of zinc, and an empirical formula was developed:

weight of carbon required = 0.655 weight of zinc produced +
0.152 weight of associated slag

From this formula a rough estimate was made of the zinc output to be expected from a charge of a given grade. With a typical high grade charge the composition of sinter fed to the furnace would be 40 per cent zinc, 24 per cent lead, and the weight of slag produced would be approximately equal to that of the zinc. Under such conditions, with a coke of relatively low reactivity, the carbon consumption required to produce 1 tonne of zinc (and 0.6 tonne of lead) would be approximately 0.8 tonne. With a lower grade charge producing more slag per unit of zinc, the production of zinc would be reduced by an amount which can be calculated from the formula. With improved standards of operation reached in recent years and with the use of higher air preheats, a number of furnaces are obtaining considerably greater zinc production per unit of carbon than would be expected from the formula, but it still has its use as an empirical guide.

The main diet for a number of smelters is formed of relatively high grade zinc and lead concentrates, supplemented by a proportion of lower grade zinciferous materials. Such furnaces as Berzelius (Germany) and Hachinohe (Japan) burn approximately 210 tonnes of carbon per day and produce annually approximately 80,000 tonnes zinc and 40,000 tonnes bullion. The weight of slag with this type of charge is comparable with that of the zinc produced. Other furnaces treat lower grade charge and whilst the carbon burning rate is similar, the zinc and lead bullion production is less, since the slag volume is greater. At Broken Hill, in Zambia, for instance, whilst some zinc and lead concentrates are treated, much of the charge is composed of residues from flotation and zinc electrolysis operations plus oxides from Waelz treatment. The sinter mix contains only 22–26 per cent zinc and 20–23 per cent lead, whilst nearly 3 tonnes of slag is produced per tonne of zinc. The annual production from such a charge is approximately 30,000 tonnes zinc and 32,000 tonnes bullion. The furnaces in Poland and in Sardinia treat considerable quantities of Waelz oxide recovered by kiln treatment of low grade ores. The Rumania furnace treats ores containing appreciable quantities of copper, which is recovered from the bullion.

Air requirements
With the exception of the larger furnace at Avonmouth (shaft area 27.1 m^2) almost all the other furnaces in operation have a shaft area of 17.1 m^2. The normal carbon burning rate of such a furnace will be 210 tonnes per day, and with an average grade charge (sinter composition 38–40 per cent zinc, 20–24 per cent lead) will produce daily 260 tonnes zinc and 130 tonnes lead. To achieve this, 36,000 m^3 (N.T.P.) of dry air will be blown per hour into the furnace bottom. The gas leaving the condenser will have an approximate analysis of nitrogen 68 per cent, carbon dioxide 14 per cent and carbon monoxide 18 per cent. Its calorific value will be of the order of 2.6 MJ/m^3 and its volume 43,000 m^3 (N.T.P.) per hour. It thus contains approximately 45 per cent of the energy of the carbon burnt, and to maintain fuel economy it is essential that as much as possible of this energy should be returned to the furnace, chiefly by using the gas from the condenser to preheat the coke and also the air blown in at the bottom.

Preheating
To preheat the air, most furnaces now use modified Cowper stoves, developed originally in the iron industry. They are essentially vertical towers packed with chequer refractory brickwork, and two stoves are usually operated in cycle, one being heated by the combustion products of condenser gas whilst the other is heating the blast air fed to the furnace. The roles of the two stoves are reversed at regular intervals to maintain a supply of hot air to the furnace.

When such stoves were first used it was expected that the refractories might be attacked by the traces of lead and zinc dust in the condenser gas, since, although this is scrubbed in towers and Theissen-type rotary washers, its dust content is still some $30-40$ mg/m^3. As a precaution, the chequerwork packing is made of high alumina brick and in practice the attack on the refractories is small.

Before stoves were first used in 1966, metal tube recuperators were used as air preheaters, but these were expensive to maintain and air temperatures exceeding 700°C could not be regularly obtained. Most furnaces now have stoves, and air temperatures have been progressively raised until now 950°C is exceeded.

So far, each increase has been profitable and on most furnaces it has been found that raising the preheat from 700°C to 900°C has resulted in an $8-12$ per cent reduction in the carbon required to produce a unit of zinc. There are indications that increases above 1000°C produce diminishing returns of useful work, and the optimum may not be far above this temperature, though the Berzelius furnace has claimed improvement at up to 1050°C preheat.

An approximate heat balance for the furnace shaft and condenser is given by Hopkin and Richards [11] and a total energy audit is given in Appendix 1.

Furnace accretions

As expected from the complexity of the reactions in the furnace — many of them being readily reversible — there is a tendency for accretion to build up on the walls. The presence of lead contributes to this in two ways. First, the lead content of the sinter fed to most furnaces is about 20 per cent and consequently relatively low melting point compounds are present, which may adhere to the walls in the upper levels of the furnace. Another cause arises from the reflux of lead sulphide in the shaft (described on page 90), but as a result of this and other reactions which result in deposition, accretions can form in the upper part of the furnace, particularly if the quality of the sinter is poor and its sulphur content exceeds 1 per cent.

Above the charge, in the furnace roof and the downcomer to the condenser, a different type of accretion occurs, from the deposition of zinc oxide by the reversal of the main reduction reaction

$$ZnO(s) + CO(g) = Zn(g) + CO_2(g)$$

This deposit is semicrystalline in character and is known with justification as 'rock oxide'. Its formation is largely held in check by the addition of top air which is blown in through ports above the furnace charge and in volume is equal to $8-12$ per cent of that of the main blast.

The gas leaves the furnace charge at a temperature varying from 850 to 900°C, depending on the furnace behaviour and the preheat temperature of the coke. It is at equilibrium and any further drop in temperature would cause deposition of zinc oxide, but with the addition of top air its temperature is raised to $1000-1020°C$. Reversion is therefore prevented in the top of the furnace and does not take place until the inlet to the condenser is reached, where cooling begins. Hence, rock oxide deposition is restricted to this zone, and to avoid excessive build-up the furnace is usually shut down for approximately eight hours at periodic $7-14$-day intervals. The oxide is removed and opportunity is taken to clean the condenser and rotors. The operation is hot and arduous, but the use of mechanical tools reduces the manual effort required.

At regular intervals — the period varying from 5 to 10 weeks — the furnace bells are taken off and any built-up accretion inside is removed by explosives. The furnace charge level is first burnt down almost to the tuyeres. Small explosive charges are detonated in steel pipes which have been driven into holes drilled into the accretion at selected points. With practice it is possible to carry out this operation expeditiously to remove any accretion almost completely, but, owing to the precautions, such a blasting session involves a furnace shut-down of $16-20$ hours. On occasions, furnaces may be blown down for complete overhaul and modification, but allowing for all stoppages a furnace is normally on full blast for approximately 86 per cent of a year.

5.4 THE CONDENSATION PROCESS

The plant

The condensers fitted to most of the standard furnaces of hearth area 17.1 m^2 in operation are rectangular in section, approximately 8.5 m long $\times$ 5.5 m wide $\times$ 1.14 m high, and contain a pool of molten lead 40 cm deep. The volume is divided by baffles into three stages, the first containing four rotors and the second and third two rotors, consisting of electrically driven vertical shafts passing through glands in the roof and terminating in a four-bladed rotor immersed in the liquid lead. The blades are inclined scoops which when rotated throw a dense spray of lead droplets filling the condenser. The gas from the furnace is forced by the baffles to take a zigzag course through this spray, and is thoroughly scrubbed, its temperature dropping rapidly from $1000°C$ on entrance to less than $500°C$ at the exit. Over 95 per cent of the zinc content of the gas is absorbed and dissolves in the lead droplets. Hot lead at a temperature of $550°C$ and containing about 2.5 per cent zinc is withdrawn from the gas inlet end and pumped into a launder where it is cooled to a temperature of $450°C$. At this temperature the solubility of zinc in lead is 2.25 per cent and the excess settles out to form a layer above the lead, which flows into a settling bath from which the zinc can be removed. The lead, at $450°C$, flows back into the gas exit end of the condenser, to be sprayed again into the gas stream.

Whilst passing through the condenser the gas simultaneously transfers heat and zinc vapour to the lead drops thrown up by the rotors. Heat is transferred by both convection and radiation, and zinc vapour by absorption, the rate being controlled by diffusion of the zinc through the gas film around each drop.

The gas stream carries the remaining 5 per cent of the zinc which is recovered as blue powder in the scrubbing system. Of this 5 per cent only one fifth is uncondensed zinc vapour, the remaining four fifths being due to a fog of zinc droplets nucleated on fume existing in the gas which enters the condenser.

The conditions under which such a fog forms can be analysed in terms of the operating lines of the system [12]. Such an analysis shows that the gas reaches its dewpoint only just prior to leaving the condenser and therefore the production of fog is reduced to small proportions.

With normal operation, using average grade charge, the gas will leave the charge at 850°C–900°C which is below the equilibrium temperature of approximately 1020°C. Air is added through the top ports, raising the gas temperature to 1040°C, which prevents reoxidation and deposition of zinc oxide until near the entrance to the condenser. At the condenser inlet a typical composition of the gas would be zinc 7 per cent, carbon dioxide 12 per cent, carbon monoxide 20 per cent. It will contain some fume (mainly lead sulphide) and dust blown over from the furnace. These have a deleterious effect providing nucleii for fog formation and producing dross, but under good operating conditions this effect is small.

The cooling launder

In the description of the condenser given above it was explained that zinc separates from lead in a cooling launder. Originally this was constructed as a deep trough with water jackets on the vertical sides. The amount of cooling could be varied by using an adjustable weir plate to control the depth of lead in the launder, and hence, the area of the cooling interface. The dimensions of the launder ensured that high Reynolds numbers, in the region of turbulent flow, were achieved at the operating lead flow rate of around 4000 tonnes per hour. A slow build-up of lead on the jackets could be removed by using baffles to increase the flow velocity.

This design has now been superseded by the use of immersible cooling coils operating in a wider refractory lined launder; this system has three major advantages, all leading to an improvement in overall condensation and separation efficiencieny:

(a) when the furnace is taken off blast, cooling is stopped by withdrawing the cooling coils rather than emptying the launder,
(b) cleaning of immersible coolers requires much less manual effort, and
(c) these coolers are more readily adaptable to automatic control.

This system is also more amenable in design to efforts to recover some of the heat dissipated in cooling water, which are now under development.

The two systems are shown diagrammatically in Fig. 24.

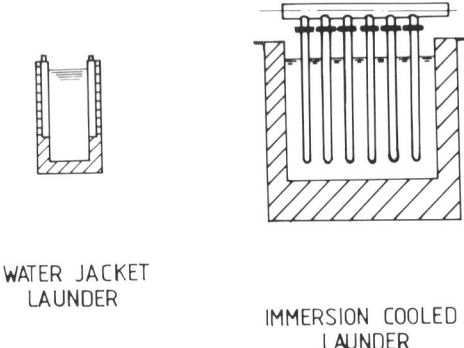

WATER JACKET
LAUNDER

IMMERSION COOLED
LAUNDER

Fig. 24 – Diagrammatic comparison of two systems of water cooling of launders
for condensation of zinc.

5.5 BEHAVIOUR OF OTHER METALS IN THE FURNACE

Lead

Lead in the sinter entering the furnace is present mainly as oxide but some sulphide, sulphate and silicate are present. The first reaction to occur in the upper zone of the furnace is the reduction of lead oxide and silicate to metal, and the bulk of the metallic lead is reduced by the carbon monoxide before reduction of zinc oxide begins. Lead sulphate is also reduced to lead sulphide and this could be further reduced by zinc vapour according to the equation

$$PbS(s) + Zn(g) = Pb(l) + ZnS(s)$$

However, as long as lead oxide is present this tends to react with the zinc sulphide to give lead sulphide, as follows:

$$ZnS(s) + PbO(s) = PbS(s) + ZnO(s)$$

In the upper levels of the furnace, therefore, whilst lead oxide is reduced rapidly, lead sulphide tends to persist, although most of the sulphur in these levels must exist as zinc sulphide which is the most stable sulphide under the conditions prevailing. As the zinc sulphide passes down the furnace and reaches the higher temperatures of the tuyere zone, lead sulphide becomes the more stable component and the reaction

$$ZnS(s) + Pb(l) = Zn(g) + PbS(g)$$

occurs, and since at these temperatures lead sulphide is volatile, it rises up the furnace. As lower temperatures are reached the reaction is reversed, with the reformation of zinc sulphide. The process is repeated and thus a reflux of sulphur in the shaft is generated.

As the molten lead descends, it dissolves and retains a number of minor metals such as silver, gold, copper, arsenic, antimony and bismuth. At the top of the shaft, the lead present restricts the volatilisation of arsenic, but a proportion of this element passes into the condenser where it causes dross formation and also dissolves in the zinc, from which it is removed by treatment with sodium as sodium arsenide. The presence of copper in the lead restricts the volatilisation of arsenic still further. The arsenic still remaining in the lead at the bottom of the furnace is converted to iron arsenides and leaves the furnace as a speiss — antimony behaves similarly, though a higher proportion leaves the furnace in the bullion.

The zinc–lead blast furnace as a lead smelter

The zinc–lead blast furnace was developed to produce zinc and this remains its most important function, but it can be considered as an efficient smelter of lead concentrates [13]. For balanced operation a minimum amount of lead must be added to compensate for that which leaves in the zinc metal and in the slag (which usually contains 0.6–1.0 per cent lead), as well as the losses which occur in the sintering and smelting operations. This minimum amount of lead (5000– 6000 tonnes per year with most operations) is an essential requirement of the process and must be regarded as a charge on the zinc operation. Frequently it is provided through the unpaid-for lead present in the zinc concentrates, and once this minimum amount of lead has been supplied, practically all additional lead added as concentrates reports in the bullion with a recovery of almost 100 per cent.

There is evidence that if the lead content of sinter added to the furnace is greater than 25–26 per cent, a tendency to accretion formation increases, and this can begin to reduce furnace throughput, but up to this limit a zinc–lead blast furnace can treat most lead concentrates more economically than they can be treated in an orthodox lead smelter. Additional lead concentrates can increase the cost of sintering, and if they increase the slag make, costs can arise through additional losses of lead and zinc. However, since most lead concentrates contain little silica this rarely happens, and frequently the slag make is reduced, because the presence of lead avoids the need to add silica to the sintering machine charge as a hardening agent. In such cases, the reduction of slag releases carbon for extra zinc reduction and the capacity of the furnace to produce zinc is actually increased, a substantial bonus being realised.

In a lead blast furnace coke is required approximating to 25 per cent of the bullion made, but in a zinc–lead furnace lead is smelted without extra coke consumption, and since almost all the capital, labour and maintenance charges are unaffected by the amount of lead treated — and are carried by the zinc operation — the operating costs due to the extra lead are almost negligible. As a result, the overall economics of the operation are improved by smelting lead concentrates, and the capacity of the furnace to do this is generally used to the maximum.

Copper

The development of the zinc—lead furnace from the original zinc furnace has also influenced policy with regards to the simultaneous smelting of copper. As in a lead blast furnace, the lead produced in the zinc—lead furnace collects copper, silver and gold introduced in the charge. As a regular feature the operating furnaces have included copper in charge up to about 5 per cent of the lead weight. On cooling of the lead bullion, the copper precipitates as a dry powdery lead—copper dross on the lead surface. The relatively low sulphur content of the charge ensures that the copper in dross is mainly metallic, whereas the dross from lead blast furnaces is sulphidic in character. Most zinc—lead blast furnace operators have sold the dross to copper refiners at a price which has given a relatively low return for the copper value, hence giving little justification for increasing the amount of copper treated.

A practical study of copper in the zinc—lead blast furnace was made by Bryson and Gray [14]. Their compilation of the results from several furnaces indicated a positive relationship between the percentage of copper dissolved in slag and the copper/lead ratio in input sinter. To some extent it might be thought that this is a natural conclusion — the copper/lead ratio in sinter dictating the copper content of lead bullion which is a liquid phase in contact with liquid slag in the furnace hearth.

More recent considerations have shown that above about 10 per cent in bullion there is likely to be little further increase in copper activity [9], so that the loss of copper in slag should not be a factor detracting from the treatment of higher quantities of copper.

Copper recovery

The poor financial return on selling copper dross stimulated the development of new processes for the recovery of copper from dross. Two hydrometallurgical processes have been developed in recent years. One of these is an oxygen—sulphuric acid leach process now in operation at the Berzelius smelter in Duisberg [15]. This produces a copper sulphate solution which is electrolysed to cathode copper and a lead—lead sulphate residue which is returned to the sinter plant.

The other process, which is operated at three smelters [16], utilises ammoniacal leaching of the dross as the first step, to form a cupric ammine

$$Cu^0 + 4NH_3 + \tfrac{1}{2}O_2 + H_2O = Cu(NH_3)_4^{2+} + 2OH^-$$

Almost all metallic ions in dross, other than copper, report in the insoluble leach residue which may be returned to the sinter plant.

The second stage of the process is concentration of the copper by solvent extraction. The copper-containing leachate is contacted with a solution of a selective organic extractant in kerosene, resulting in the ion-exchange transfer of copper ions to the organic phase. The aqueous raffinate, depleted in copper, is recycled to the leach stage for re-use. The organic phase loaded with copper is scrubbed with weak acid to remove trace impurities and then contacted with recirculating copper sulphate—sulphuric acid solution to strip the copper from the organic to the aqueous phase by the reverse ion-exchange reaction of the loading stage.

The products are a 'stripped' extractant solution which is recirculated to the load stage and a pregnant copper sulphate solution from which either high grade cathode copper or pure copper sulphate may be obtained for sale.

A simple flowsheet of the process is given in Fig. 25.

These copper recovery processes have underlined the advantages to be gained in increasing the throughput of copper in the zinc—lead blast furnace even though the amounts are low compared with the outputs of zinc and lead. Technical advances in furnace and drossing practice will need to be achieved before plants can operate consistently with greater than 10 per cent copper in bullion.

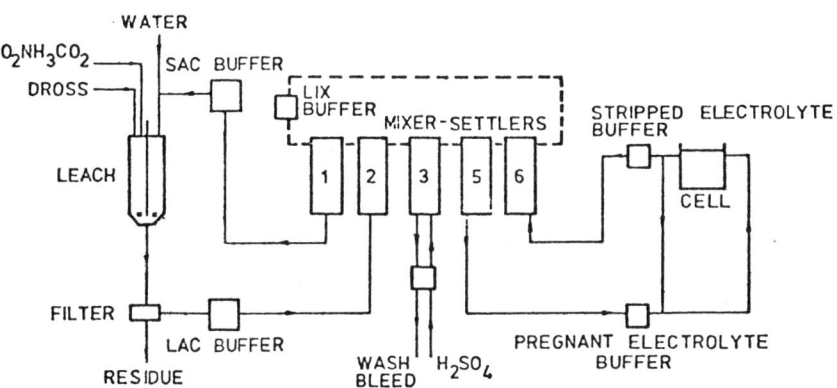

Fig. 25 — Flowsheet of process for copper recovery from copper dross.

Tin

Tin is normally present in charge only at low levels, but it is mostly volatilised, and, since even small concentrations of tin (0.02 per cent) in zinc are associated with embrittlement, the permitted tin content in the feed on installations in which at least part of the metal produced is sold as Grade 4 metal has to be restricted. The situation is different if the whole of the zinc metal produced is refined, as the tin values can be recovered and add to the revenue from the treatment.

Treatment of complex materials and residues

Most zinc—lead blast furnaces were established to process primary zinc and lead concentrates, but the ability of the process to accept a wider range of materials is being increasingly exploited with the incorporation of complex mixed ores and residues from other processes; and the wide range of mixed feed currently processed in the Sulphide Corporation smelter at Cockle Creek, Australia, and the Berzelius smelter at Duisberg, West Germany, has been reviewed in detail [17]. At Cockle Creek, low grade concentrates, dump retreatment concentrates and other process residues constitute about 50 per cent of the smelting load, and as they contain sulphur they are sinter-roasted with high grade concentrates. One of the inputs is a silver—lead leach residue from the Electrolytic Zinc Company of Australasia Limited's jarosite process — an example of the blast furnace consuming an environmentally difficult residue generated by the other major zinc-producing process.

At Duisberg a wide range of zinc and lead oxidic residues are included in the furnace feed. These include zinc ashes and skimmings, lead smelter dusts, steel works dust high in lead and zinc and Waelz oxides produced from low grade residues. To enable these materials to be utilised without resource to sintering, a special briquetting technique is used.

Briquetting techniques for furnace feed

The advantage of a briquetting process is the possibility of incorporating oxidic materials into furnace feed, by-passing the sinter—roasting plant. In the late 1960s a ram press of 3 tonne/h capacity was installed at Duisberg for the cold-briquetting of zinc ashes arising from the galvanizing industry. Annually this unit provided 10,000 tonnes of compressed ashes, containing 6000 tonnes of zinc. In the mid-1970s, research into cold-briquetting of zinc/lead dross arisings and coke fines was studied at the Australian plant at Cockle Creek [18]. Laboratory-Scale briquetting led to the successful development of a 2 tonne/h briquetting pilot plant which now operates as a production unit fully integrated with the smelter complex. The process is relatively simple and involves the use of a special binder. Earlier work was proceeding in the Research Department of Imperial Smelting Processes on a process of hot-briquetting of oxidic materials in a roll press in the absence of binder. A successful technique was developed which resulted in the commissioning in 1975 of a large-capacity hot-briquetting plant at the Berzelius smelter. Today this plant provides about 25 per cent of the zinc input to that ISF (at least 20,000 tonnes/annum zinc). Hot-briquetting plants have also been installed at the Samim plant in Sardinia and the Sumitomo plant in Japan.

A flow sheet of a typical briquetting operation is shown in Fig. 26.

Use of tonnage oxygen

Developments in chemical engineering over the last 30 years have improved the efficiency of tonnage oxygen production to such an extent that its use as a reagent in commercial processes is now relatively widespread, particularly in the steel industry. Operating units are now available which produce oxygen either as a gas of a purity exceeding 95 per cent, or as enriched air, with a power consumption of less than 500 kWh per tonne.

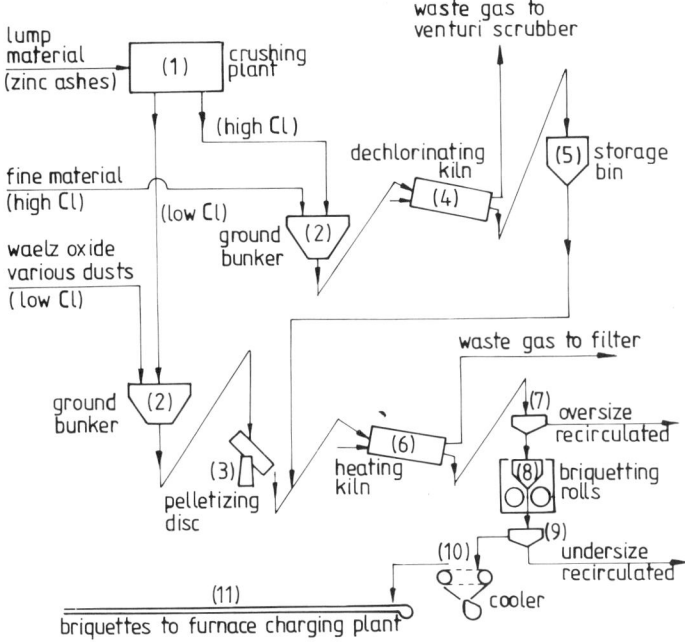

Fig. 26 – Flowsheet of hot-briquetting plant at Duisburg.

At first sight it would appear that the use of tonnage oxygen to enrich the air blown through the tuyeres of a zinc–lead blast furnace might be of value, since it could be used to raise the zinc content of the gas entering the condenser. The efficiency of the lead splash condenser operating on a standard air-blown furnace (zinc content of gas 7 per cent) is already so high, however, that the effect of this aspect of enrichment is negligible.

It would appear that the only major benefit from oxygen enrichment in zinc–lead blast furnace practice would arise in a case where it was desired to increase the capacity of an existing furnace which was already being blown as hard as possible. Enrichment would then enable the zinc content of the gas produced in the furnace to be raised. Even in such cases, oxygen enrichment should only be considered after equipment to give the maximum preheat possible has been installed. A trial using oxygen additions to raise the content of the blast to 28 per cent was carried out in 1963 on the Swansea furnace, which largely confirmed these conclusions.

With a new installation, the capacity to produce the output desired should be provided from the start, and enrichment would not then be required.

Two processes using tonnage oxygen have been proposed which, if fully developed commercially, might have some impact on the future of zinc metallurgy. The Kivcet process — a Russian development — treats mixed lead–copper–zinc sulphide concentrates. These are burnt in suspension, in a flame using oxygen, in a shaft or cyclone. By adjusting the conditions, a slag is formed containing most of the zinc and lead as oxides, together with metallic lead, and a matte phase containing the copper. The two latter phases can be removed and subsequently processed, leaving the zinc oxide-containing slag to be treated in a vessel attached to the main reaction shaft. This vessel is fitted with electrodes, which heat the bath and reduce and volatilize the zinc. It is claimed this can be recovered either by passing it through a lead-splash type condenser, or by burning it to form oxide, which can be recovered electrolytically. The first commercial unit to be built outside the USSR is at Potosi in Bolivia.

The main advantages that the process offers are the avoidance of a roasting or sintering stage, and that it does not require a supply of metallurgical coke. It is essentially a lead smelting process, however, and would not appear to be a rival to the electrolytic or blast furnace processes for major zinc recovery.

The QSL process (named after the inventors P. E. Queneau, R. Schuhmann and Lurgi, the developing company) also uses tonnage oxygen, and does not require metallurgical coke. Pellets of sulphide concentrates are fed into a molten slag bath held in a rotating cylindrical furnace. Oxygen is fed into the bath below the surface, reacting with the sulphur present, forming sulphur dioxide and generating heat. Further along the pool, pulverised coal is added, which reduces the lead oxide contained in the slag, leaving the zinc oxide present largely unreduced. Molten lead and a copper matte can be recovered as separate liquid phases. Zinc oxide can be recovered from the slag by fuming or other methods.

The process would appear to hold potential advantages over present orthodox lead smelting practice, but is unlikely to effect existing zinc production techniques.

5.6 APPLICATION AND ECONOMICS

As a result of the intensive development work carried out at Avonmouth, the first fully commercial zinc blast furnace unit was built at Swansea and came on line in March 1960. In 1984 thirteen furnaces were in operation producing approximately 733,000 tonnes of zinc and 368,000 tonnes of lead annually — roughly 12 per cent and 8 per cent of world production. The process has established itself as the best of the thermal methods, but in recent years most new plants have been based on the electrolytic process. The reasons for this are not clear-cut, as will be seen from an analysis of the operating costs of the two processes.

Comparative cost of electrolytic and blast furnace processes

Under present conditions of growing inflation, rising energy charges and rapidly changing exchange rates, it is not possible to give a meaningful estimate of the capital cost of a new plant of either type. Site conditions are all-important. In the *Mineral Commodity Profile on Zinc 1983* published by the US Bureau of Mines, some figures for construction costs of electrolytic plants are quoted from which Table 9 has been constructed.

Fig. 27 – The No. 4 zinc blast furnace at Avonmouth. (Courtesy of Commonwealth Avonmouth Company.)

Table 9

Capital cost of recently built electrolytic plants

		Annual production zinc (tonnes)	Capital cost per annual tonne
1982	Industrial Minera Mexicana San Luis Potosi	113,000	$1150
1981	Catamrouilla Peru	100,000	$3200
Estimate not yet built	Noranda Mines New Brunswick	100,000	$3000
1984	Thailand	60,000	$2400

Much of the equipment and infra-structure making up the major part of the capital cost is common to both processes. Each requires roasting, sulphuric acid and cadmium plants, together with ore-handling facilities, access to road and rail, power and water supplies, workshops and change rooms. A blast furnace plant requires a refluxing unit to make high grade metal, and extensive ventilation equipment; but an electrolytic unit must spend more on residue despatch. On the whole it would appear that the capital cost of a blast furnace complex is generally less than that required for an electrolytic unit of equal capacity, but the difference is not likely to be great. The differences are summarised in Table 10.

However, operating costs are only a part of the overall profitability, which is largely determined by the cost of the raw materials treated. The feed for practically all electrolytic plants consists almost entirely of relatively high grade concentrates, and if the impurity content of the feed varies, operating difficulties can develop quickly. The blast furnace is less sensitive to charge grade, and can treat a range of lower grade materials. At Broken Hill in Zambia, whilst some lead and zinc concentrates are treated, much of the charge is composed of residues from flotation and electrolytic operations. The furnaces at Katowice, Poland, and Porto Vesme, Sardinia, treat considerable quantities of Waelz oxide, recovered by kiln treatment of low grade ores. The Rumanian furnace treats ore containing considerable quantities of copper, which is recovered from the bullion.

At some mines (Woodlawn mines in New South Wales is a recent example) a bulk zinc–lead concentrate is made, giving benefit in increased recovery of values, and such material forms an excellent feed to a blast furnace. Such practice is not widespread, however, owing probably to an unenlightened buying policy adopted by blast furnace operators.

Table 10

Summary comparison of electrolytic and blast furnace processes

	Electrolytic	Blast furnace
Plants included	Roasting Sulphuric acid Leaching Purification Electrolysis Cadmium production	Roasting Sulphuric acid Blast furnace Zinc refinery Lead refinery Cadmium production
Output (tonnes per annum)	Zinc 100,000–120,000 Grade 1 Sulphuric acid 170,000– 200,000 Cadmium 200–240	Zinc 100,000–120,000 Grade 1 Lead 50,000 Sulphuric acid 200,000– 220,000 Cadmium 220–240
Labour, including supervision and all services, man hours per tonne	6–8	9–10
Energy required per tonne zinc	Power 4200–4400 kWh Steam 2 tonnes net Fuel oil 738.5 MJ	Coke (dry) 1.1–1.2 tonne Power 580 kWh Fuel oil 3165 MJ
Maintenance (annual charge expressed as percentage of capital costs)	5–6	6–7

An important recent development is that most furnaces now supplement their feed to the maximum with secondary materials such as brass residues (from which both zinc and copper can be recovered) galvaniser ashes, and flue dusts from the steel industry. This is valuable from the viewpoint of conservation and the environment, and since in such materials the price of the contained zinc has, of necessity, to be below that in high grade concentrates, blast furnace profitability is improved.

At Duisberg in Germany, at Samim in Sardinia, and at Sumitomo in Japan, Waelz kilns are used to upgrade zinciferous materials (containing <20 per cent zinc). The zinc oxide so produced is hot-briquetted after pelletization, at temperatures of 700–750°C. Good bonding is obtained, and since the material contains little sulphur, the briquettes can be charged directly to the furnace, and form a suitable feed.

Overall recovery

As far as overall recovery of zinc is concerned, blast furnaces attain 92–94 per cent, unless lower grade material forms a large part of the feed. Most electrolytic plants treating high grade concentrates, with jarosite-type precipitation, on residue treatment, claim an overall zinc recovery of 94–96 per cent.

A considerable advantage held by the electrolytic process comes from the development of fluidised bed roasting. Roasting costs are low with the large fluid bed furnaces now widely used, with their high capacity and low maintenance and operating costs. The recovery of approximately 1 tonne of steam for every tonne of blende roasted is an added bonus. The blast furnace is forced to adopt sintering, owing to the presence of lead and the need to produce lump material. This is an expensive operation with high maintenance costs, and no system exists to recover the heat liberated in the roasting operation.

An apparent advantage of the electrolytic route is that it produces metal of >99.95 per cent grade, and if required can produce 99.99 per cent zinc at little extra cost. The blast furnace produces GOB grade (1.2 per cent lead) and though there is a considerable market for this grade, particularly in general galvanising, an additional refluxing stage must be used to make high grade zinc. To give a balanced view of the two processes, however, it must be recognised that the electrolytic process can operate only with very high purity zinc sulphate, hence the cost of complete 'refining' of all production is a necessity.

As is dealt with in detail in Chapter 9, hygiene and pollution pose special problems. With the electrolytic process there is, in general, less difficulty in avoiding atmospheric pollution, but the disposal of liquid effluents and jarosite and other precipitates is more onerous than with the blast furnace wastes. As the blast furnace is a zinc and lead smelter, the stringent specifications for limiting lead contamination must be satisfied, both inside and outside the plant. On the other hand, the main waste product – the furnace slag – is stable and can be stored in dumps without risk of contaminating surface water. Electrolytic plants do not avoid entirely the complication of handling lead, since they produce a lead residue in a fine state of division, which has to be shipped to a smelter for treatment. At both plants, extreme care has to be taken to avoid pollution of any kind and to work within the existing hygiene standards.

To summarise, it seems likely that new zinc production in the immediate future will rely on the electrolytic or blast furnace routes, the choice depending on a number of factors of which costs of production form only a part. The relative availability and cost of electric power and coke are obviously of major importance. If the smelter is to produce zinc alone, with no lead production, and if adequate supplies of suitable concentrates and electric power are available, the electrolytic route would be preferred. On the other hand, if the production of both lead and zinc is required, and if there is access to mixed concentrates and secondary materials, then the blast furnace is superior.

REFERENCES

[1] Derham, L. J., British Patent 572961.
[2] Woods, S. E., British Patent 682179.
[3] Morgan, S. W. K., The production of zinc in a blast furnace, *Transactions of the Institution of Mining and Metallurgy,* Vol. 66, pp. 553–65, 1956–7.
[4] Sellwood, R. M., *World Symposium on Mining and Metallurgy of Lead and Zinc, AIME, New York,* p. 581, 1970.
[5] Harris, C. F., *International Blast Furnace Hearth and Raceway Symposium, Newcastle, Australia,* Australian IMM, March 1981.
[6] Gammon, M. W., *Symposium on Advances in Extractive Metallurgy, IMM, London,* pp. 47–52, 1977.
[7] Lumsden, J., The physical chemistry of the zinc blast furnace, *Proceedings of the Metallurgical Chemistry Symposium, Brunel University and the National Physical Laboratory,* July 1971, Paper 44, pp. 533–48, HMSO, London, 1972.
[8] Barin, I. and Knacke, O., *Thermochemical Properties of Inorganic Substances,* Springer-Verlag, 1973.
[9] Richards, A. W., *Canadian Met Quarterly,* Vol. 20(2), pp. 145–151, 1981.
[10] Harris, C. F., Richards, A. W. and Robson, A. W., Lead–zinc–tin '80, pp. 246–260, *AIME, Las Vegas, 1980.*
[11] Hopkin, W. and Richards, A. W., *J Metals,* Vol. 30(11), pp. 12–17, 1978.
[12] Morgan, S. W. K. and Woods, S. E., Application of the blast furnace to zinc smelting, *Metallurgical Review,* Vol. 16, p. 157, 1970.
[13] Morgan, S. W. K. and Greenwood, D. A., The metallurgical and economic behaviour of lead in the Imperial Smelting Furnace, *Journal of Metals,* Vol. 20, December, 1968.
[14] Bryson, J. L. and Gray, P. M. J., Recovery of copper in the Imperial Smelting Furnace, *Transactions of the Institution of Mining and Metallurgy,* Vol. 77, pp. 72–84, 1968.
[15] Tack, K., *Erzmetall,* Vol. 29, pp. 276–279, 1976.
[16] Hopkin, W., Hunter, W. H., Nakade, K., Asano, A. and Kobayashi, S., *Symposium on Hydrometallurgy AIME, Atlanta, Georgia,* pp. 985, 1983.
[17] Adami, A. O., Firkin, G. R. and Robson, A. W., Complex metallurgy '78, *Joint GDMG-IMM Symposium, Bad Hargburg,* pp. 36–42, 1978.

[18] Tavener, M. G. and Buchanan, A. S., *Symposium: Agglomeration '77, AIME*, pp. 737–753.

6

The electrolytic process

The electrolytic process has made great progress since commercial operation started in 1917. Its spectacular growth, particularly during the last twenty years, has been due largely to the application of physical chemistry (especially to the problem of iron precipitation), the utilisation of advances in chemical engineering, and the recently developed rapid methods of analysis and control.

At the moment, its only rival for new installations is the zinc–lead blast furnace, and for the treatment of high grade zinc concentrates it holds a number of advantages over the thermal method. Perhaps the greatest of these is that it has, of necessity, to produce metal of high purity directly, whereas the blast furnace produces low grade metal, which, to satisfy the high grade markets, must be refined by refluxing.

6.1 BACKGROUND

Some large electrolytic units have recently been built. The largest is the Valleyfield plant of Canadian Electrolytic Zinc Limited, which, it is claimed, will reach an output of 225,000 tonnes per annum. The plant of Kempensche Zinc Maatschappiz at Budel (see Fig. 33), which commenced operation in 1974, has a rated capacity of 150,000 tonnes per year, subsequently increased to 180,000 tonnes per year. Most of the other new plants have a capacity of 100,000 tonnes per year or less. By comparison a standard zinc blast furnace with a shaft area of 17.2 m² is capable of producing 80,000 tonnes per annum of zinc and 40,000 tonnes of lead. The large furnace (shaft area 27.1 m²) built at Avonmouth has produced 90,000 tonnes of zinc and 37,000 tonnes of lead annually, but it is generally believed that the furnace itself is capable of producing 130,000 tonnes of zinc a year and 60,000 tonnes of lead, if increases to the ancillary equipment (sintering and acid plant) are made. Both systems therefore can gain the benefits of large unit production.

A comparison of the energy balances, capital costs and operating requirements of the electrolytic and blast furnace processes has been given in Chapter 5.

Before the beginning of this century numerous attempts were made to develop a hydrometallurgical method of producing zinc based on electrolysis. In 1881 Létrange patented a method covering the electrolysis of sulphate solutions but made no commercial application of his work. In the 1890s E. A. Ashcroft proposed to treat rich zinc–lead ore from the Broken Hill field in Australia. A plant was built at Cockle Creek near Newcastle, New South Wales, where, after roasting of zinc concentrates, the zinc oxide was leached, and the zinc was recovered from the solutions by electrolysis. Both chloride and sulphate solutions were used at various times, but the project was unsuccessful owing mainly to deposition difficulties [1]. From 1910 to 1915 Brunner Mond operated a chloride electrolysis process in which zinc oxide was heated in the calcium chloride solutions discarded from the Solvay process. Carbon dioxide passed into the slurry precipitated calcium carbonate, and the zinc chloride formed was purified and then electrolysed using revolving iron discs as cathodes, with carbon anodes. Chlorine formed at the anodes was collected and converted into bleaching powder, but owing to inefficient deposition and corrosion troubles, the process was abandoned. Attempts were made to produce zinc by electrolysis of molten zinc chloride, but these again were unsuccessful, mainly because of the difficulty of producing oxygen-free melts.

In 1910 interest was aroused again in the possibility of sulphate electrolysis, originally suggested by Létrange. Work by the Anaconda Company, the Consolidated Mining and Smelting Company of Trail, and the Electrolytic Zinc Company of Australasia at Risdon showed that the main problem was to obtain adequate solution purity. Zinc is electronegative to hydrogen and deposition of the metal is possible only if the overvoltage, or resistance to liberation of hydrogen, is high, which is so in the almost complete absence of impurities — particularly antimony, arsenic, cobalt and germanium. It was thus necessary to develop techniques for purifying the sulphate electrolyte to extreme limits before continuous deposition of zinc could be assured.

The first commercial plant at Anaconda, Montana, reached a daily capacity of 25 tonnes in 1915. In 1916 larger plants were in operation at Great Falls, Montana, and at Trail, followed in 1918 by a plant at Risdon, Tasmania. The process has since grown steadily and today is the most widely applied method of zinc extraction, producing at least 75 per cent of the world's zinc.

6.2 ELECTROLYTIC PROCESS OPERATION

The main feed to all electrolyte plants is zinc sulphide concentrates, and in almost all cases the first step is to roast the blende, converting the zinc sulphide into zinc oxide, and evolving sulphur dioxide which is converted into sulphuric acid. For the roasting operation the majority of plants use fluosolid roasters, the operation of which is described in Chapter 3.

Cell operation and deposition efficiency

In the electrolytic process as practised today a solution of zinc sulphate is electrolysed between lead alloy anodes and aluminium cathodes. Zinc is deposited on the cathodes and is periodically removed. Sulphuric acid is formed at the anodes and oxygen evolved. The acid is circulated outside the cells, and dissolves fresh zinc oxide forming zinc sulphate which is then electrolysed.

The success of the process depends essentially on the reactions occurring at the cathodes. An analysis of the reactions showing the overwhelming importance of hydrogen overvoltage was developed by Bratt and is given in detail in Appendix 3.

The power required for electrolysis is the product of cell voltage and current. Cell voltage is the sum of a number of terms the most important of which are the reversible cell voltage, the electrolyte resistance, the oxygen overvoltage, and the anode scale voltage drop. An analysis of these voltage components of an operating cell at the Risdon plant is given by Gordon [2]. This shows that the combined reversible electrolytic potential under operating conditions is 2.036 volts, some 59 per cent of the cell voltage.

Individual effect of impurities

The effect and method of removal of some of the most important impurities are summarised below.

Magnesium, sodium, potassium

The sulphates of these elements are not decomposed in the cell, and therefore they have little influence on electrolysis. If they build up in solution a portion must be discarded after precipitation of the zinc. Levels of sodium and potassium can be controlled to some extent by jarosite precipitation as described later in this Chapter.

Calcium

Calcium is another element which plays no direct part in the electrolysis. Calcium sulphate has only limited solubility, but can cause difficulty owing to deposition in pipes and launders carrying the solutions and in the cooling equipment.

Aluminium

Aluminium has little direct influence on electrolysis, but can have a beneficial effect in forming complexes with fluorides present, thus reducing the tendency to form adherent cathode deposits.

Iron

Iron should be removed during the early stages of purification. If present in quantity (its content is usually kept below 10 mg per litre in the electrolyte) it causes a reduction in current efficiency since it tends to be reduced at the cathodes and oxidised at the anodes.

Manganese
Manganese is present in most zinc ores, but its overall effect in solution is not harmful. A proportion is deposited as manganese dioxide on the anode as a sludge, and this is beneficial, since it tends to reduce corrosion of the anode, which can in turn lead to deposition of lead on the cathode causing contamination. Some of the sludge is retained as a scale on the cathode, which increases cell voltage, but some falls to the bottom of the cell also as a sludge and the rate at which this occurs determines the frequency of cleaning of the cells and anodes. The precipitated manganese dioxide also plays a useful part in removing, by adsorption, a proportion of the copper, arsenic, antimony and cobalt present and thus eases the load on purification. The manganese content of electrolytes is generally held below 7g/l although operation at higher levels has been practised.

Cadmium, copper, lead
Copper must be removed, since it reduces current efficiency and also contaminates the zinc. Cadmium and lead, although not reducing current efficiency, also cause contamination of the zinc. Removal of cadmium in the zinc-dust purification stage presents little difficulty, although care must be taken not to allow the precipitated cadmium to stay in contact with zinc sulphate, otherwise re-solution occurs. Copper can be completely removed also at the zinc-dust stage, but precautions must be taken to prevent subsequent contamination of the electrolyte by corrosion products from copper contacts. Lead may occasionally be present as a result of insufficient zinc-dust treatment but more usually arises from corrosion of the anodes. The addition of silver to the anodes reduces this effect. Strontium carbonate or barium hydroxide are sometimes added to lower the lead content of the solution by coprecipitation of the lead and strontium (or barium) sulphates.

Cobalt
If present in excess, cobalt tends to cause perforation of the zinc deposits, and can prevent electrolysis entirely, although small concentrations reduce anode corrosion and can thus be beneficial. Its effect can be partially controlled by glue or other additions. Its influence on the process is critical and its content is carefully monitored. Most plants reduce the cobalt content to below 1 mg/l, but at some plants solutions containing 10 times this concentration are used. In order to remove cobalt, additions of potassium antimony tartrate, with the zinc dust used as a precipitant, may be made, but the more usual practice is to precipitate cobalt by an excess of zinc dust in the presence of copper and arsenite compounds, at temperatures above 70°C. Great care must be taken to prevent the generation of arsine — an acute poison. Alternatively, cobalt may be precipitated as cobalt-nitroso-beta-naphthol by the addition of sodium nitrite, beta naphthol and sulphuric acid, after iron and copper have been removed.

Nickel

Nickel, particularly in combination with cobalt and other impurities, can be harmful when present in concentrations above 0.1 mg/l, causing the production of holes in the zinc deposits. The necessary degree of removal is generally reached, without difficulty, during the normal zinc dust precipitation stages.

Antimony

Antimony is one of the most harmful impurities and can prevent deposition entirely. It cannot be tolerated above 0.2 mg/l, and its harmful effect is increased if cobalt or germanium are present.

Arsenic

Arsenic is less harmful than antimony and can be tolerated at concentrations up to 1.0 mg/l. It can be removed efficiently at the iron precipitation stage and most plants run very much below this figure.

Germanium

Germanium is present in some zinc ores, and has a very harmful effect in electrolysis, particularly in association with cobalt and antimony, as it causes widespread production of holes in the cathode deposits. It can rarely be tolerated above 0.01 mg/l, and its removal can be complicated. A proportion is removed during iron precipitation and the remainder with the zinc dust, but sufficient time is necessary for complete precipitation to occur. If the germanium concentration is sufficiently high a separate removal stage may be introduced, in which a precipitate of germanium tannate may be formed, and this can be treated subsequently to recover germanium.

Chlorine

Chlorine forms perchloric acid at the anode and can cause severe corrosion there, whilst at high concentrations chlorine gas evolved at the anode may become troublesome to cell operators. Concentration of chlorine up to 0.4 g/l may be tolerated, and the excess can be removed by adding silver sulphate, but this is expensive unless recovery of the precipitated chloride is complete.

Fluorine

Fluorine is troublesome, since if present above 0.2 mg/l it can cause sticking of the zinc to the aluminium cathode and can thus interfere with the stripping operation. It is claimed that the effect of fluorine can be controlled by the addition of aluminium sulphate which forms aluminium fluoride complexes. Fluorine is partly removed during neutralisation by adsorption on iron and aluminium hydroxide.

Conductivity

Apart from the above considerations, elements present in significant amount affect the resistivity of the solutions. The higher the concentration of salts present, the higher the viscosity and hence the resistivity, with a consequent effect on cell voltage. Elements such as sodium, potassium, magnesium, manganese and of course zinc, are all important in this respect. The resistivity of solutions is also materially affected by the acidity, an increase in which increases the conductivity.

The necessity to reach and maintain a high standard of electrolyte purity dominates the process. The development of rapid and continuous methods of analysis has ensured that the process can be controlled at all stages. Variations in composition of the incoming calcines can be quickly detected, and the necessary adjustments made to purification procedure to ensure that the electrolyte can be maintained at a high level of consistency, and a steady level of production obtained.

6.3 PRE-ELECTROLYSIS PROCESSING

Although a number of improvements have been introduced since the process was originally developed by Anaconda and at Trail, the basic flowsheet remains essentially unchanged. This is shown in Fig. 28.

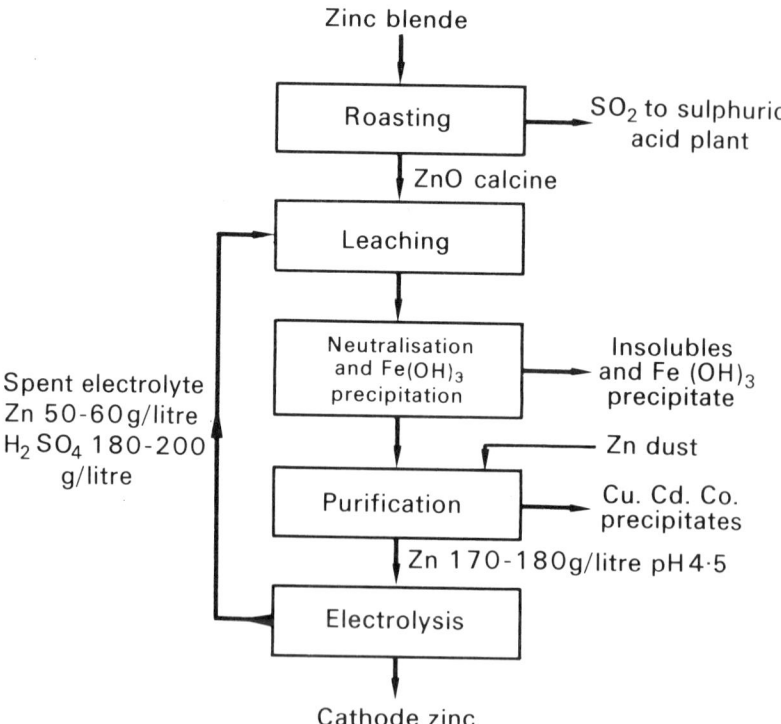

Fig. 28 – Basic flowsheet of electrolytic process.

Leaching

The main function of the leaching stage is to dissolve the zinc oxide from the calcines, using sulphuric acid regenerated in the electrolytic cells. This is combined with oxidation and neutralisation, so that iron can be precipitated carrying with it a number of impurities such as arsenic, antimony and germanium. Colloidal silica is coagulated and precipitated, as is aluminium hydroxide.

Operation differs from plant to plant: some use batch leaching, but most modern plants leach and purify continuously. Although at some plants both leaching and precipitation are carried out in a single stage, it is today more usual to employ two stages (Fig. 29). With one-stage leaching, calcine is added to return acid, and the soluble zinc is dissolved. As originally developed, the strength of sulphuric acid in the initial leaching solution was of the order of 100–150 g/l, and this readily dissolved the zinc oxide in the calcines, no external heat being required. Under these conditions zinc ferrite, $ZnOFe_2O_3$, formed during roasting is not dissolved, but is left in the residues, and unless recovered by subsequent treatment this zinc is lost. When all available zinc oxide had been dissolved, the solution was ready for precipitation of the iron.

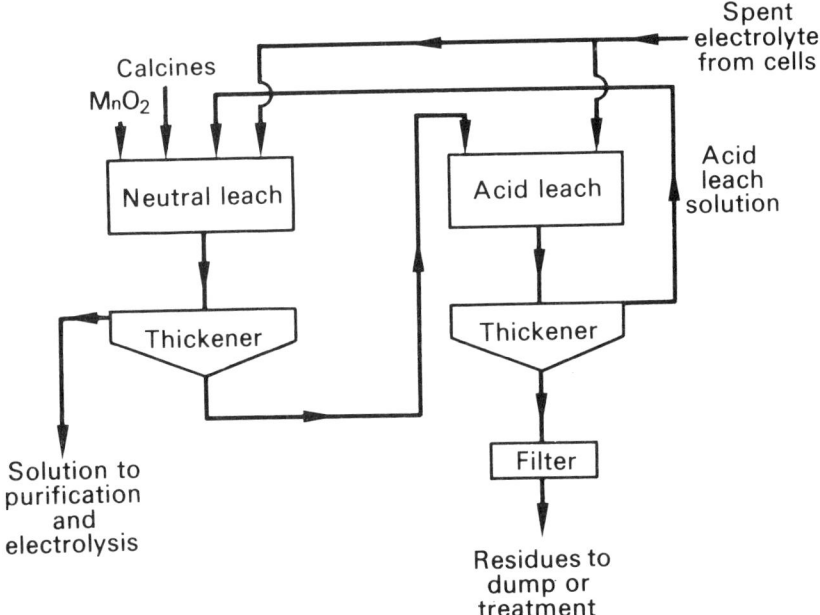

Fig. 29 – Two-stage leaching circuit.

At this stage, iron is present mainly as ferrous iron. Whereas the precipitation of ferric ions begins at pH 3.5 and is complete at pH 5, ferrous ions remain in solution. Therefore oxidation is essential to ensure iron removal, and although sometimes manganese dioxide is added to assist oxidation, it can be completed using air alone. Good agitation and dissemination of air are necessary to obtain a high rate of oxygen solution. Study of the rate of oxidation of ferrous oxide in zinc sulphate solutions according to the reaction

$$2Fe^{2+} + 2ZnO + 3H_2O + O = 2Fe(OH)_3 + 2Zn$$

shows that it does not occur below pH 1.9, but in more neutral solutions the rate increases exponentially with pH in the range 2–5.5. Thus, to obtain the necessary rate of oxidation, sufficient calcine must be added to effect neutralisation to pH 5.5. Under these conditions reasonably rapid oxidation of the ferrous iron occurs, with precipitation of a filterable ferric hydroxide. One of the first detailed studies of the mechanism of iron precipitation was published by Posnjak and Merwin [3].

The first step of the two-stage process is called the neutral leach. Excess calcine is added to approximately half the return cell acid, together with the cell solution from the second stage, to a pH of 5.0–5.2, and ferric iron precipitated with most other impurities. After settling, the neutral solution is sent forward to the final purification section. The residue, containing approximately 50 per cent of the zinc, is leached with acid in the second stage and the soluble zinc dissolved. After settling, the still-acid solution is passed to the first-stage leach.

Under the conditions outlined above, silica is coagulated and precipitated, as is aluminium hydroxide; the iron content is reduced to less than 0.01 g/l and almost all the arsenic, antimony and germanium are removed. The mechanism of this removal is not entirely clear, but it has been attributed to the formation of insoluble basic arsenates and similar compounds, and to adsorption involving iron, aluminium and silica. Whilst the precipitation of iron as ferric hydroxide was used for many years, it has now been displaced, and at most plants today the bulk of the iron is precipitated either as jarosite (usually $(NH_4)_2Fe_6(SO_4)_4(OH)_{12}$), goethite (FeOOH) or as hematite (Fe_2O_3).

Zinc-dust purification

After the leaching and iron removal stages, the composition of the neutral solution can be expected to be within the limits shown in Table 11.

Using zinc dust, the next step is to precipitate copper, cobalt and cadmium and to reduce arsenic, antimony and germanium to acceptably low levels. The purification is generally carried out continuously in several stages, in a series of agitated tanks, to yield solutions of composition indicated by Table 11.

Purification is effected by zinc-dust precipitation, relying on the fact that all the elements to be removed lie below zinc in the electrochemical series and can therefore be precipitated by cementation. The presence of arsenic, copper or antimony activates the zinc dust, increasing its selectivity and efficiency. The practice at most plants is first to precipitate cobalt at 90 °C at a pH of approximately 4. If sufficient arsenic and copper are not present in the original solution, copper sulphate and arsenious oxides are added. Copper, nickel, arsenic and

Table 11

Zinc-dust purification — typical composition ranges of neutral and purified solutions

	Neutral solution pH 4.5–5.2	Purified solution pH 4.5–5.5
Zinc	100–180 g/l	110–180 g/l
Manganese	2.5–20 g/l	2.5–20 g/l
Cadmium	10–500 mg/l	0.1–0.5 mg/l
Copper	10–600 mg/l	0.02–0.5 mg/l
Iron	1–10 mg/l	1–10 mg/l
Cobalt	2–20 mg/l	0.4–1.0 mg/l
Nickel	1–10 mg/l	0.05–0.1 mg/l
Arsenic	0.1–0.5 mg/l	0.1–0.2 mg/l
Antimony	2–100 mg/l	0.04–0.2 mg/l
Germanium	10–100 mg/l	0.001–0.01 mg/l

antimony are precipitated at this stage. After removal of the precipitate, the solution passes to the second stage where the addition of more zinc dust removes cadmium and thallium. This stage is generally carried out at pH 3 and 70–80°C. These procedures are usually adequate for the removal of other impurities such as germanium.

Whilst the use of arsenic as an activator of the zinc dust is efficient, and relatively cheap, it introduces toxic risks, and at Vielle-Montagne and other plants the addition of arsenic has been replaced by that of antimony salts. Using antimony additions, the first purification stage is generally carried out at 65–75°C, and most of the copper, nickel and cobalt, together with some of the cadmium, is precipitated. The filtered solution is then cooled and treated again with excess zinc dust to remove the remainder of the cadmium and any of the other metals still remaining.

In a number of plants recently built, the stages have been reversed. The solution is first treated in the cold with zinc dust, to precipitate cadmium and copper. The liquor is then heated to 90°C and cobalt and germanium removed by a second zinc-dust addition in the presence of antimony. This procedure enables purer precipitates to be made and increases the recovery of the metals in them.

At Porto Maghera and some other plants [4] where the copper content is low (0.09 g/l) and the cobalt is relatively high (0.011 g/l), a separate cobalt precipitation is carried out. Following a first stage with zinc-dust additions, acid is added to bring the solution to pH 2.4 and a solution of sodium beta napthol and sodium nitrite added to precipitate the cobalt as cobalti-nitroso-beta napthol. This is removed by filtration and the cobalt-bearing product recovered for sale. Excess reagent is removed by addition of activated carbon, and a third stage of zinc-dust addition follows, to remove the remainder of the copper and any germanium or antimony. The solution is then cooled to 70°C and additions of zinc dust sufficient to precipitate the cadmium are made. This again is removed by filtration and recovered, but care must be taken to avoid oxidation of the cadmium sponge otherwise some may be redissolved.

Zinc dust, due to its fine state of division, is a useful reagent, and is used in almost all plants for electrolyte purification. The amount used varies with the content of impurities to be removed and with the current density employed, but can be as much as 6 per cent of the cathode zinc production, although the stoichiometric demand is only of the order of 0.5–1.5 per cent. This inefficiency arises mainly from the fact that a proportion of the zinc particles become coated with layers of precipitated metals, immobilising the zinc below. Claims are made that improved utilisation efficiency can be obtained if precipitation is carried out in vibrated beds of zinc granules, but large-scale experience of the suggestion has not yet been published.

6.4 ELECTROLYSIS

After the purification stages, and provided analysis shows that the content of all relevant impurities has been reduced to the necessary levels, the solution is passed forward for electrolysis, which is carried out in a series of rectangular cells, usually constructed of PVC sheets supported by a steel frame. The cells are often laid out in a series of cascades with the electrolyte flowing continuously from one to the other (see Fig. 30). The number of cells in each cascade varies from 3 to 15.

The number of anodes and cathodes per cell also varies from plant to plant. In the lower range, 20–25 cathodes may be used per cell, whereas other plants may use as many as 48. The cathodes are arranged at a fixed spacing, with the centre–centre distance between homopolar electrodes usually 90 mm. They are connected electrically together, as they rest on a bus-bar. They can be removed in groups for stripping and replacing by means of overhead cranes. The anodes are similarly placed at fixed spacing between the cathodes and rest on a positive bus-bar, and can also be removed together for cleaning.

The cathodes are made of aluminium sheet, usually argon arc welded to aluminium header bars, which support them, and electrical contact is made with the bus-bar via a copper contact piece. The cathodes are generally about 1000 mm long, 600 mm wide, and 5 mm thick, thus providing about 1.2 m^2 of surface for deposition. Recently at Balen, Vieille-Montagne have developed cells using cathodes with a deposition surface of 3.2 m^2. The use of these 'jumbo' cathodes, with a completely automated system of handling and stripping,

Fig. 30 – Cell room at Budelco, Holland.

represents a considerable advance, with a considerable reduction in labour and capital costs [5]. Several other plants are now using jumbo cathodes. Cominco at Trail Canada have completely replaced their old plant with a modern automated plant using 3 m^2 cathodes, and having a capacity of 270,000 tonnes of zinc per year – the largest in the world [6].

The anodes, which are slightly smaller than the cathodes they serve, are made of cast or rolled lead sheet, to which alloying additions of silver are made as required. This important addition (varying from 0.5 to 1.0 per cent silver) reduces anode corrosion, and therefore contamination by lead of the cathode deposit. The anodes are frequently perforated with circular holes, said to assist circulation of the electrolyte and also to help maintain a protective film of manganese dioxide on the surface.

The operation of the cells requires the removal of the considerable amount of heat liberated, which depends upon the current density used, the composition (and therefore conductivity) of the electrolyte, and a number of other factors. Under normal conditions it amounts to 3.5 to 4.0 GJ per tonne of cathode zinc produced, and accounts for some 30–35 per cent of the energy fed to the cells. In most plants today this heat is removed by passing a proportion of the acid solution leaving the cells to vacuum evaporation or to cooling towers, and returning the cooled liquor to the cells. The ratio of recycled cooled solution to new feed may vary from 4 to 1 to as much as 15 to 1. The temperature of the cell electrolyte is generally maintained in the range 30–40°C, the usual preference being 35–38°C.

There is considerable variation in the acid and zinc levels in the electrolyte used in various plants, and in the amount of zinc stripped during passage through the cells. The choice of zinc and acid content in the solution leaving the cells depends upon the impurity level in the feed solution and the acceptable current efficiency. The amount of zinc removed in each pass ranges from 50 to 75 per cent, with a corresponding variation in the acid content of the solution.

Electrical supply

In order to make full use of the considerable amount of power required, an electrolytic plant demands a high standard of electrical engineering. To effect the conversion from alternating to the direct current used in the cells, silicon rectifiers are widely used, and conversion efficiencies of over 97 per cent are obtained.

The current density used is determined by the length of the stripping cycle chosen. For convenience of labour utilisation this is usually either of 24 or 48 hours' duration, and may vary from 8 to 72 hours, but with the growing use of automatic stripping techniques this is becoming of less importance. With 48-hour stripping, current densities may vary from 400–500 A/m^2, but shorter times are required if higher current densities are used. Cell voltage, determined by current density and the conductivity of the solution, varies from 3.3 to 3.5 V per cell, while current efficiency varies from 88 to 92 per cent. The consumption of energy during electrolysis lies generally within the range 3100–3300 kWh per tonne of cathode zinc produced.

The considerable rise in the cost of energy over recent years has forced many plants to adopt a flexible approach to their energy demand over the 24 hours, reducing consumption (by lowering current density) at high cost peak load conditions, and increasing it during cheaper off-peak periods. Thus current density and production rate can vary from hour to hour, and skill is required to adjust other conditions to compensate for the fluctuations.

Cleaning of electrolyte

Since manganese is present in almost all zinc ores and is not removed during the purification stages, it forms one of the main impurity constituents in the electrolyte. It plays little part in the electrodeposition process at the cathode, and can be tolerated in concentrations up to 25 g/l. It is partly deposited as manganese dioxide on the anodes, forming a scale, and within limits this is beneficial since it protects the anode from corrosion. As growth continues however it becomes excessive, and an increase in cell voltage occurs. Some of this growth drops from the anode and forms a sludge which builds up in the bottom of the cell, and therefore both the anodes and the cell must be cleaned periodically to remove excess manganese dioxide. Calcium sulphate, as gypsum, together with silica, is also slowly deposited, mainly in the launders and cooling equipment, which also require regular cleaning.

Cathode stripping

At the end of the deposition period, depending on the stripping cycle chosen, the cathodes are removed from the cells to a stripping bay where the deposited zinc is stripped or peeled from them. They are then repaired if necessary and returned to the cells and electrolysis continued.

With correct conditions, stripping presents little difficulty as the zinc separates readily from the aluminium sheet on which it has formed. If purification has been inadequate, the zinc may adhere firmly and stripping becomes laborious. The presence of excess fluorine in solution is one of the most frequent causes of sticking.

With the availability of modern methods of rapid chemical analysis it is now possible to control conditions of purification and of electrolysis so that serious sticking no longer occurs. This has permitted the development of mechanisation, pioneered at Vielle-Montagne [6]. A number of plants are now using systems in which the operations of cathode removal, transfer to stripping stations, stripping of zinc and return of stripped cathodes to the cells are all carried out mechanically and automatically.

Melting of cathodes

After stripping, the zinc cathodes are usually washed, dried and then melted. This is most frequently carried out in low frequency induction furnaces, which vary in capacity, the largest having inputs of up to 1800 kW. Power consumption is of the order of 100–120 kWh per tonne of zinc cast. Dross production varies from 2 to 2.5 per cent, and thus the recovery of cast zinc is 97.5–98 per cent, with the dross usually being returned to the circuit via the roaster.

The molten metal is fed into casting machines and cast into slabs for sale, this operation, including the stacking of the slabs, usually being carried out automatically. Part of the production may be cast into large blocks of about one tonne mass. These suit modern large-scale alloying and galvanising plants.

6.5 MAJOR DEVELOPMENTS

The basic process has been described, but important developments have taken place. Modern practice still relies on the essential principles developed by the pioneers at Trail and Anaconda, and the considerable advances made since that early work are due more to improvements in technique and operation than to radical change in the process.

Low lead zinc

Prior to 1926—7 the grade of zinc produced by electrolytic plants was running at better than 99.95 per cent zinc, the main impurities being lead 0.02—0.03 per cent, cadmium 0.01—0.02 per cent and iron 0.01 per cent. This was considerably purer than any made by the horizontal retort or any other pyrometallurgical process at the time.

A series of zinc alloys had been put forward containing 4—10 per cent aluminium and 1—2 per cent copper, for the newly developed die-casting industry. These appeared to have good mechanical and casting properties, and had considerable application, but after some months of service a serious defect became apparent. Rapid deterioration in physical properties occurred, and in humid atmospheres the casting sometimes collapsed completely.

The New Jersey Zinc Company showed that this catastrophic effect was due to the presence of small quantities of low-melting-point impurities, such as lead, tin, bismuth and, to a lesser extent, cadmium. It was proved conclusively that if the total sum of impurities in the alloy was reduced to less than 0.01 per cent, stable and satisfactory castings could be produced. This was a major discovery, but at the time the electrolytic zinc industry was unable to produce in quantity metal of the purity required.

U. C. Tainton solved the dilemma [7] when he showed that anode corrosion, which was the main reason for the presence of lead in the cathode zinc, could be greatly reduced by alloying the lead anodes with 1 per cent of silver. By casting lead—silver anodes and by improving the efficiency of cadmium and iron removal, it was demonstrated at the Bunker Hill plant of the then Sullivan Mining Company that zinc containing less than 0.01 per cent of total impurities could be made. The metal thus produced was of sufficient purity to make stable die-casting alloys, and this opened up one of the main markets for zinc.

Lead—silver anodes are now always used in modern practice when zinc of the highest purity is required. To control the lead content even further, the practice of adding strontium or barium salts in conjunction with the use of lead—silver anodes has been developed.[8]

6.6 PROBLEMS OF ZINC FERRITE FORMATION

In the early days of electrolytic zinc production roasting was mainly carried out in circular-hearth-type furnaces. An advance was made by the development by Cominco at Trail of the flash roasting process, but these methods have now been displaced by fluid bed roasting as developed primarily by Dorr Oliver and Vielle-Montagne. The advantages of this process are low labour and maintenance costs, high sulphur dioxide content of the flue gas, and good heat recovery in the form of steam. Most electrolytic plants now use some type of fluid bed roasting, but whilst benefitting from the above advantages, difficulties due to zince ferrite $ZnO.F_2O_3$ production have increased. Whilst in hearth roasting some ferrite was formed, particularly in the presence of marmatite, in which ferrous sulphide occurs isomorphously with zinc sulphide in the ore, it was generally possible, to some extent, to limit the iron combining with zinc oxide by controlling the hearth temperature. In suspension and fluidised bed roasting, where temperatures in excess of $900°C$ are maintained, conversion of practically all the iron oxide to zinc ferrite occurs, the effect being more marked in fluidised bed roasting possibly because of the relatively long roasting time.

In what have been called standard or basic electrolytic plants (although few plants today operate in this manner), the calcine is leached with return spent electrolyte and no heat is applied other than that generated in the reaction. Temperatures in the leaching vats do not exceed $60°C$, and in these conditions all free zinc oxide present in the calcine is readily dissolved, but zinc ferrite remains virtually insoluble. In a way, this is an advantage since it reduces the amount of iron taken into solution, which would require subsequent precipitation as ferric hydroxide. This is bulky and difficult to wash free from zinc salts. Thus, though the insolubility of zinc ferrite in the standard leaching procedure reduces the amount of iron to be handled, it causes a serious loss of zinc in the residues, unless these can be subsequently treated. Hence, recovery of zinc in the standard process, in the absence of any residue treatment, rarely exceeds 88—90 per cent, and this was a serious handicap to the development of the process.

A considerable amount of work on the problem was carried out by the industry and a number of solutions proposed and tested. The most promising line of attack was to dissolve the zinc ferrite in the residues in hot sulphuric-acid-containing solutions, and then to tackle the problems of removing iron from the solutions so obtained.

6.7 THE JAROSITE PROCESS

The first major breakthrough came with the development of the 'jarosite' method of precipitation. An insoluble basic iron sulphate mineral of approximate formula $K_2Fe_6(OH)_{12}(SO_4)_4$ had been first recognised at Jaroso in Spain and had been given the name 'jarosite'. Work at a number of laboratories showed that insoluble jarosite-type precipitates could be formed during neutralisation of ferric sulphate solutions in the presence of complexing agents. Such precipitates were crystalline in character and had higher density and settling and filtering rates than orthodox ferric hydroxide. A number of metals were found to act as complexing agents, the most suitable being sodium, potassium, or ammonium (NH_4).

Industrial application of this discovery soon followed. Simultaneously, but independently, work at the Electrolytic Zinc Company at Risdon, Det Norske Zinkkcompanie in Norway, and Asturiana de Zinc SA in Spain, showed that, with typical plant liquors, 90–95 per cent of the iron content could be precipitated as a jarosite with good settling and filtering characteristics, by adding the necessary sodium or ammonium ions and adjusting the pH to 1.5. The remaining iron in solution could then be removed as ferric hydroxide. More iron in solution could thus be tolerated, and leaching techniques could be used which extracted ferrite as well as zinc oxide, thus increasing recovery of zinc.

The development of jarosite precipitation is relatively new, and its full potential has not yet been demonstrated. It is applied at Norzink, Electrolytic Zinc and Asturiana SA where the method was initially developed, but also at a number of other plants [9].

Its development has enabled a number of variations of the basic flow sheet to be made, primarily with the aim of improving the recovery of zinc and also of lead–silver and other values present. The simplest means of incorporating the new precipitation method into a plant flow sheet is shown in Fig. 31.

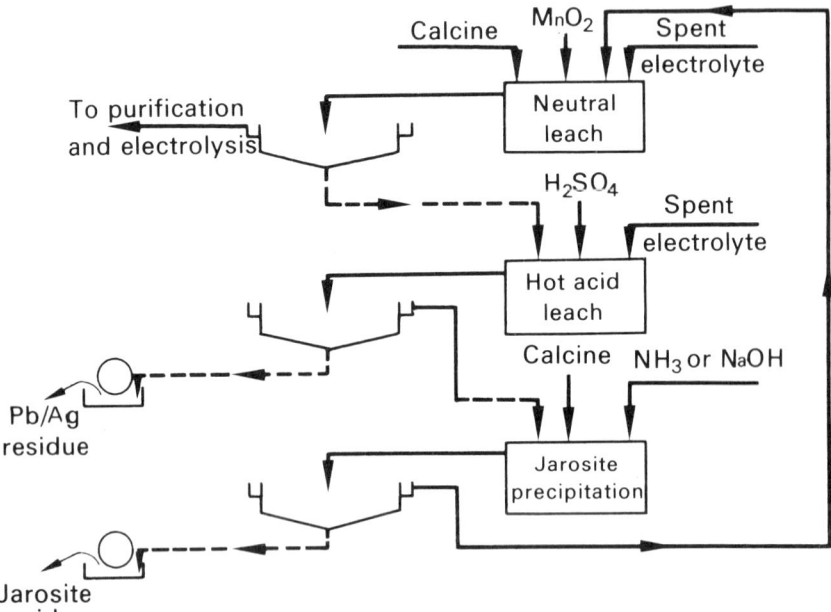

Fig. 31 – Simplified flowsheet of the hot leach and jarosite precipitation process.

In the first or neutral leach stage, calcine is leached with spent electrolyte and the zinc oxide is dissolved but not the zinc ferrite. Some manganese dioxide is usually added to assist oxidation of the ferrous iron, and neutralisation continues until pH 5 is reached and ferric hydroxide is precipitated. When this reaction is completed the slurry passes to a thickener. The clarified overflow is pumped to the zinc dust purification stage prior to electrolysis. The thickened residue from the neutral leach passes forward to the acid leach where it is attacked with spent electrolyte to which sulphuric acid is added, and the temperature raised to 90–95°C. Under these conditions ferric hydroxide is redissolved and zinc ferrite is attacked.

At the end of this stage more of the zinc and iron are in solution. The insoluble residue, now containing most of the lead and silver originally present in the ore, is removed by settling and filtration. Some value for it can generally be obtained from a lead smelter or a zinc–lead blast furnace.

After removal of the lead–silver residues and other insolubles, the solution is passed forward to the jarosite precipitation stage. If necessary the acidity is reduced to pH 1–1.5; then ammonia is added and precipitation of jarosite begins, or, instead of ammonia, sodium ions can be added as sodium hydroxide or sodium sulphate, depending on cost and availability. During precipitation of the yellow jarosite the temperature is maintained in the range 90–95°C.

As will be seen from the equation

$$3Fe_2(SO_4)_3 + 10H_2O + 2NH_4OH = (NH_4)_2 \, Fe_6(SO_4)_4(OH)_{12} + 5H_2SO_4$$
$$\text{ammonium jarosite}$$

sulphuric acid is liberated during the reaction, and as precipitation proceeds, zinc oxide, usually in the form of calcine, is added to maintain the pH at the desired value.

When precipitation is complete, the slurry passes to a thickener where the jarosite settles and can be removed. The solution may still contain some iron, mainly in the ferrous state, and this is removed either by oxidation and neutralisation in a separate stage or by return to the neutral leach.

Whilst this represents what may be termed the original process, a number of refinements are practised to improve zinc recovery. Thus, the jarosite precipitate can be washed with dilute sulphuric acid in which ferrite is soluble but jarosite is not. This reduces loss of zinc in the jarosite since the solution can be returned to the circuit. The ferric hydroxide precipitated in the final neutralisation stage can be returned to the acid leaching section, but this is not always possible, since its content of arsenic, antimony and germanium may be high.

As described on page 123 one major advantage of jarosite precipitation, which is already exploited, is that it is now possible to treat old residues produced before the new method was applied, and which still contain a considerable quantity of zinc as zinc ferrite. These residues can be fed to the hot acid leaching stage and most of the zinc ferrite dissolved. Jarosite precipitation copes successfully with the additional iron taken into solution. At the same time lead, silver and cadmium left in the old residues can be recovered. Thus, with the development of jarosite precipitation, zinc ferrite need no longer be left in the residues but can be extracted and recovered. Previously, due to the loss as ferrite, recovery of zinc from the orthodox electrolytic process rarely exceeded 88 per cent but now with suitable use of jarosite precipitation recoveries of 96 per cent are possible.

Jarosite precipitation gives a further advantage. In the roasting operation some formation of zinc sulphate cannot be avoided and build up of sulphate in the circuit can occur. This can be an embarrassment, and to hold this in check at a number of plants some solution must be periodically withdrawn and discarded, after recovery of its zinc content. Since jarosites are essentially basic sulphates their removal from the circuit provides a useful bleed-off of this radical, and usually no additional discard of solution is required.

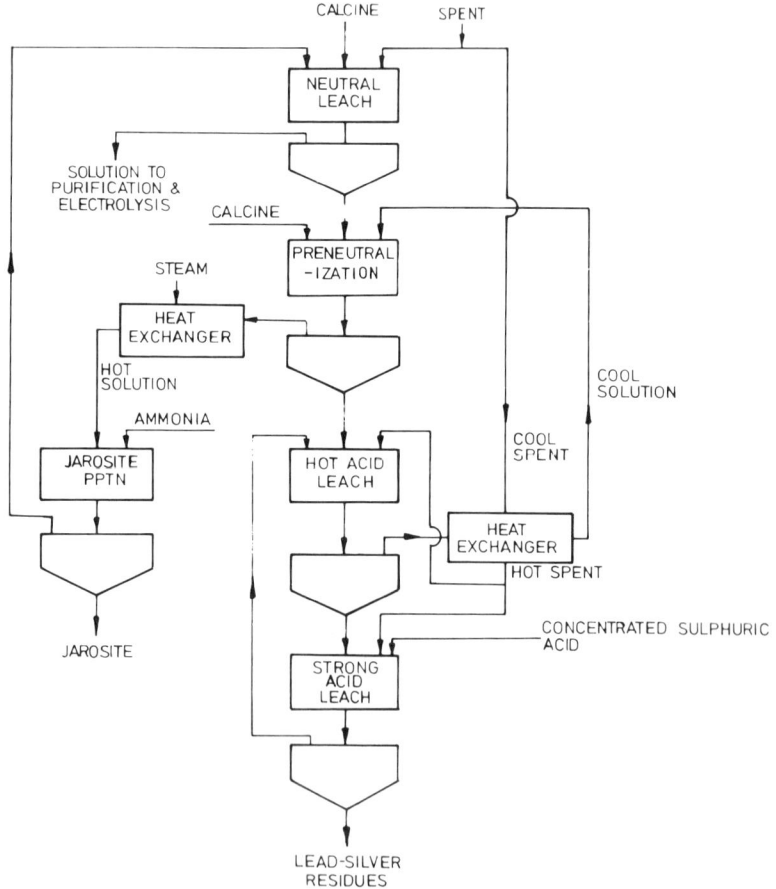

Fig. 32 — Improved jarosite process.

Further development of the jarosite process has continued. The Electrolytic Zinc Company of Australasia Limited has tested new procedures in which jarosite is precipitated without the use of a neutralising agent in the jarosite precipitation step [10]. The most convenient way of achieving this is shown in Fig. 32.

The hot acid leaching step is conducted at 90–95°C. Solution from this step is cooled to 50–60°C, and then neutralised with calcine. At the lower temperature, jarosite does not form very readily or rapidly and during the short time required for neutralisation, virtually no jarosite forms. After neutralisation, residue is separated and returned to hot acid leaching and a solution obtained which is high in iron but very close to neutrality. This solution is heated to 90–95°C at which temperature jarosite forms readily at the low acidity. The jarosite precipitation eventually ceases because of the increase in acid concentration caused by the acid released by the jarosite reaction. However, 70–80 per cent of the iron present is precipitated as pure jarosite. The residual solution proceeds to neutral leaching where the remaining iron is precipitated and eventually returned to hot acid leaching where it is re-dissolved and returned to jarosite precipitation, so that all iron is ultimately precipitated as jarosite.

This new process has the important advantage that no calcine is used for neutralisation in the jarosite step so that no zinc ferrite or silver and lead compounds from the calcine are lost in jarosite. Accordingly the recovery of zinc is increased from 96–97 per cent to near 99 per cent. Recovery of lead and silver in the hot acid leach residue is increased from, say, 70 per cent to better than 95 per cent.

This improved procedure has been extensively and successfully tested in a large pilot plant at the Electrolytic Zinc Company of Australasia Limited's Risdon plant and no doubt industrial application will follow in due course.

6.8 OTHER PRECIPITATION PROCESSES

Goethite precipitation

An alternative method of precipitating iron in crystalline form was developed at the Balen plant of S.A. Vieille-Montagne where a number of improvements to zinc metallurgy have been made in the past [11]. In the Balen process, calcine is heated as in standard practice and the residues then retreated with hot acid solutions in which zinc ferrite is attacked. The ferric iron dissolved from the ferrite is first reduced to ferrous iron, using zinc sulphide as the reducing agent:

$$Fe_2(SO_4)_3 + ZnS = 2FeSO_4 + ZnSO_4 + S$$

The sulphur-containing residue from this operation is returned to the roasting plant. The ferrous iron is then reoxidised to the ferric state using air or oxygen:

$$2FeSO_4 + O + 3H_2O = 2FeOOH + 2H_2SO_4$$

Simultaneously as the ferric iron is formed, it is precipitated from solution as goethite, FeOOH, although some β-FeOOH and α-Fe$_2$O$_3$ are also present. The reaction is carried out at pH 3–3.5 for 4–6 hours at 90°C, in either batch or continuous fashion. Calcine is added progressively as the reaction proceeds, to counter the acidity developed. Under correct conditions the goethite is mainly crystalline and can be filtered and washed without great difficulty.

As far as cost and performance are concerned there seems little to choose between the jarosite and goethite methods of precipitation but the former is probably somewhat easier to control and operate. Both types of precipitate are in a fine state of division and contain small quantities (approximately 1 per cent) of soluble zinc. With the growing concern about pollution, the storage of the considerable quantities produced annually presents problems. They must be retained at most sites in large-scale pond systems so that all seepage can be collected and returned to the plant for removal of toxic impurities. In this respect goethite has some advantage in that its iron content (42–43 per cent) is greater than that of jarosite (25–30 per cent), and therefore the bulk produced for a given zinc output is somewhat less, but the disposal problem by either route can be considerable.

A description of the development of jarosite and goethite precipitation and the effect on electrolytic plant practice is given by Gordon and Pickering [9].

Haematite precipitation – Akita developments

A possible method of circumventing some of the environmental problems associated with jarosite or goethite residue disposal is in use at the Akita Zinc Company's plant in Japan [12]. At Akita the filter cake of leach residues is repulped with spent electrolyte, and then passed at a temperature of 100°C through brick-lined autoclaves, pressurised at 207 kPa with sulphur dioxide, and under these conditions most of the iron, zinc, copper, cadmium, etc., can be dissolved. The solution is then treated with hydrogen sulphide, and a precipitate containing most of the copper, silver and lead is removed by filtration. From this precipitate, by using flotation, a concentrate can be made containing most of the copper and silver, which can be treated by copper smelters. Limestone is then added to the filtrate in two stages, the first, to pH 2, producing a precipitate of calcium sulphate, which is marketable as gypsum. The second stage to pH 4.5, precipitates almost all the other impurities, including gallium, germanium, arsenic and antimony. In the final stage of the process the filtrate is treated with oxygen in titanium-lined autoclaves, at a temperature of 200°C and 2070 kPa, under which conditions the iron is precipitated as haematite, with an iron content of approximately 60 per cent. The iron content is therefore sufficiently high to justify treatment in an iron blast furnace, but as the precipitated material contains some zinc, sulphur and arsenic, treatment to remove these impurities is generally necessary before acceptance by iron smelters.

A variant of the haematite precipitation method as used at Akita has been developed at the Datteln plant of the Ruhr-Zink company. Owing to environmental considerations, at Datteln dumping of jarosite or goethite residues was not possible, and it was decided to adopt the haematite precipitate method to eliminate iron in a form which could be disposed of commercially. Since at Datteln they did not wish to be involved in calcium sulphate production as gypsum (as at Akita), a decision was taken to use zinc blende as neutralising agent. The zinc/iron solution from hot acid leaching is consequently treated at high temperature (but normal pressure) with zinc blende. At this stage most of impurities are precipitated and iron is reduced to the ferrous state. From the now-neutral solution the iron is precipitated as haematite by oxygen in autoclaves heated to 180°C at a pressure of 15 bar (1.5 MPa) [13].

6.9 LEACHING

All electrolytic units built recently, and most of the older plants, use a jarosite, goethite or haematite method of iron removal and the residues finally produced contain little zinc.

Treatment of leach residues

Until these new methods of precipitation were developed, removal of iron as ferric hydroxide was standard practice, and since such precipitates were bulky and difficult to wash free from occluded zinc sulphate, it was essential to restrict the amount of iron taken into solution. For this reason, leaching conditions were chosen in which free zinc oxide was dissolved but zinc ferrite left unattacked, and at most plants large quantities of residues accumulated containing considerable quantities of zinc and iron, and appreciable amounts of other values. For some time little attempt was made to treat these residues (although they could contain up to 15 per cent of the zinc received in the ore), but many techniques were later developed and applied. At first, pyrometallurgical methods were used, employing heated rotary kilns, relying on the fact that at temperatures above 900°C, in the presence of carbon, the zinc oxide in zinc ferrite is reduced to metallic zinc. This leaves the bed as a vapour, to be reoxidised again in the kiln atmosphere above the bed. The zinc oxide so formed leaves the kiln as a fume which, after cooling, can be caught in a bag filter. Most of the lead is also volatilised and can be retained with the zinc oxide.

As jarosite precipitation became established, a number of plants commenced the treatment of old residues by leaching them with hot acid solutions, in which zinc ferrite is soluble. After precipitation of the iron as jarosite, a zinc sulphate solution is obtained which can be purified and electrolysed. As would be expected from one of the originators of the jarosite process, this application has been carried furthest at the Risdon plant, in Tasmania, of the Electrolytic Zinc Company of Australasia Limited, where a plant has been built to treat approximately 70,000 tonnes of current and 50,000 tonnes of stockpile residues per year. The residue plant is producing 23,000 tonnes of zinc per year and 220 tonnes of cadmium, as well as saleable copper and lead residues. The overall recovery from the residue plant is 80 per cent, raising the zinc recovery from the whole plant to 94 per cent.

Direct leaching of concentrates

As described earlier in this chapter, the first stage at almost all electrolytic plants is to roast the basic feed of zinc blende in furnaces of the fluosolid type, forming zinc oxide, which can be readily leached in the return cell liquor. Roasting involves the production of sulphur dioxide which must be converted on site into sulphuric acid. Difficulties can arise in disposing of all the sulphuric acid product (roughly equal in weight to that of the blende treated). As a consequence, Sherritt Gordon Mines Limited have developed a process, which they used originally on nickel sulphide ores, in which blende is treated with sulphuric acid (spent cell liquor) in autoclaves with oxygen under pressure at 150°C. The basic reaction is:

$$ZnS + H_2SO_4 + \tfrac{1}{2}O_2 \ = \ ZnSO_4 + S + H_2O$$

The zinc sulphate solution is passed forward for standard electrolytic plant treatment, and the sulphur can be recovered and stored, or shipped for treatment elsewhere. The process thus avoids the necessity to produce and dispose of immediately, large quantities of sulphuric acid stoichiometrically equivalent to the zinc production. Where this presents no problem, it is doubtful whether pressure leaching will displace roasting, but in certain cases the greater flexibility it gives can be of conxiderable value. The process has been installed by Cominco Limited at Trail [14] and will be installed at Kidd Creek Mines at Timmins.

6.10 ENVIRONMENTAL CONSIDERATIONS

The rapidly growing world-wide awareness of the dangers of atmospheric and effluent pollution has had a major effect on the industry, and is now one of the main factors influencing the techniques employed. To comply with the already very strict limits in impurity control of effluents, which all plants today have to observe, virtually all heavy metals must be retained and no escape can be tolerated. Since any contamination of surface water must be avoided, material can no longer be stored in dumps if it contains water soluble impurities, unless measures are taken to retain and treat all drainage. Most plants today have a final effluent treatment section in which all liquors leaving the plant are treated (mainly with lime) and the toxic elements precipitated to the limits required. At many plants the final effluent is potable and can sustain fish. The standards of effluent control which most companies adopt and the methods in general use to attain them are dealt with more fully in Chapter 9.

Fig. 33 — General view of Budelco electrolytic plant in Holland.

6.11 RECOVERY OF CADMIUM

At most electrolytic zinc plants the main bulk of the cadmium is precipitated in the second zinc-dust stage after the precipitation of copper in the first. After settling and filtration, a cake is obtained which lies generally within the limits:

Zn	25–30%
Cd	10–15%
Cu	5–10%
Pb	2–4%

The cake also contains small amounts of antimony and arsenic and sometimes nickel and cobalt. As with zinc, before cadmium can be produced by electrolysis, impurities must be removed almost completely. Zinc can be tolerated, but the zinc–cadmium ratio in the electrolyte should not exceed 1:2.

To achieve these conditions the first stage in cadmium extraction is to redissolve the cake in dilute sulphuric acid (spent cell liquor is frequently used). Cadmium and zinc in the cake are dissolved, but copper and lead are left in a residue which can be sold to copper smelters. In the early days of cadmium extraction a number of fatalities occurred as a result of arsine poisoning, and great care is taken to ventilate all vessels containing cadmium sponge, especially during leaching operations.

After the cadmium has been redissolved, a second zinc-dust precipitation follows, and a reasonably pure cadmium sponge is obtained, which is redissolved in sulphuric acid and, if sufficiently pure, the solution is passed to the electrolytic cell. If thallium is present at this last stage, the solution is neutralised to pH 4.5 and the thallium precipitated with sodium bichromate.

Cells used for the electrolysis of cadmium are usually similar to those used in the main zinc plant. The composition of electrolyte fed to the cells varies from plant to plant and lies generally within the limits:

Cd   100–200 g/l
Zn   40–100 g/l

The acid content in spent cell solution lies generally in the range 100–170 g/l. Cell temperature is kept as low as possible and should not be allowed to exceed 35°C or re-solution from the cathodes becomes serious. Current density used generally lies between 60 and 100 A/m^2, with current efficiency between 85–88 per cent and deposition times varying from 16 to 72 hours.

After deposition, the cathodes are removed from the cells, stripped, melted under caustic soda and cast into the shapes the market requires. Owing to the toxicity of cadmium, care must be taken to protect the operators from any fumes arising from the molten metal.

At most plants the cadmium produced is of high purity and a typical analysis would be:

Cd   99.99%
Pb   0.0011%
Zn   0.001%
Th   0.00005%
Cu   0.00007%

REFERENCES

[1] Ashcroft, E. A., The electrolytic deposition of zinc, *Trans. Inst. Min. Metall,* Vol. 6, pp. 282–337, 1897–98.

[2] Gordon, A. R., Improved use of raw material, human energy resources in the extraction of zinc, *Proc. Advances in Extraction Metallurgy 1977,* London, April 1977. Institution of Mining and Metallurgy.

[3] Posnjak, E. and Merwin, H. E., *Journal of American Chemical Society,* Vol. 44, p. 1965, 1922.

[4] *World Symposium on the Mining and Metallurgy of Lead and Zinc,* American Institute of Mining and Metallurgical Engineering, Vol. 2, p. 284, 1970.

[5] de Bellefroid, Y. and Delvaux, R., New Vieille–Montagne cell house at V. M. Balen Plant, IBID p. 204.

[6] Jansen, H., New electrolytic and melting plant, *Australian Mining,* Feb., 1984.

[7] Tainton, U. C., Taylor, A. G. and Ehrlinger, H. P., *Transactions of the American Institute of Mining and Metallurgical Engineers*, Feb., p. 16, 1929.

[8] Bratt, G. C. and Smith, W. N., The effects of strontium compounds in the electrolytic production of zinc, *First Australian Conference on Electro-chemistry*, Sydney and Hobart, Feb. 1963.

[9] Gordon, A. R. and Pickering, R. W., *103rd American Institute of Mining and Metallurgical Engineers Annual Meeting*, Dallas, 24–28 Feb. 1974.

[10] Matthew, I. G., Haigh, C. J., Pammenter, R. V., Low contaminant jarosite process — pilot plant results, *Annual Meeting American Institute of Mining and Metallurgical Engineers*, Los Angeles, Feb. 1984.

[11] Andre, J. A. and Masson, N. I. J., The goethite process in retreating zinc leaching residues, *American Institute of Mining and Metallurgical Engineers*, Dallas, 24–28 Feb. 1974.

[12] Tsunoda, S., Maechiro, I., Emi, E. and Sekine, K., Electrolytic zinc plant of Akita Zinc Co., *Annual Meeting American Institute of Mining and Metallurgical Engineers*, Paper A73–65, Feb. 1973.

[13] Ropenack, A., Hematite — the solution of a disposal problem. An example from the zinc industry, *Erzmetall*, Vol. 35, No. 10, p. 537, 1982.

[14] Veltman, H. and Bolton, G. L., Direct pressure leaching of zinc blende with simultaneous production of elemental sulphur, *Gesellschaft Deutscher Metallhutten und Bergleute, E. V. Annual Meeting*, Berlin, Sept. 1979.

7

Thermal refining of zinc

Before the development of the electrolytic process in 1917 all zinc was produced in horizontal retorts and gave typical analyses around the following:

lead	1.3%
iron	0.03%
cadmium	0.2%
zinc	98.5%

Although metal of this grade was adequate to satisfy the then main applications of hot-dip galvanising and rolled sheet, demands for purer zinc were growing, particularly in the brass industry. The electrolytic process produced much purer metal (zinc purer than 99.95 per cent), but when the New Jersey Zinc Company in the late 1920s showed that stable zinc-base die-casting alloys could be produced only when the quantity of impurities present was kept at a level which demanded metal containing over 99.99 per cent zinc, the industry could not at first supply the grade required. This problem was soon overcome by the New Jersey Zinc Company which developed the refluxing process, primarily for the refining of vertical retort metal, and also through the work of U. C. Tainton and others [1] on the electrolytic process which enabled metal of the necessary high purity to be obtained, largely through the use of lead anodes alloyed with silver.

7.1 SPECIFICATIONS

In recent years the tendency has grown for consumers of zinc to adopt tighter specifications for the metal they purchase, and even when alloy additions have to be made, zinc of high purity is usually demanded. The major exception is, in general, hot-dip galvanising where Grade IV zinc (lead <1.35 per cent, iron <0.04 per cent, and zinc not less than 98.5 per cent) is satisfactory in most cases, since high lead content is usually beneficial. For continuous strip galvanising lines a debased high grade zinc alloy containing 0.15 per cent lead and 0.55 per cent aluminium is often specified.

Chapter 12 describes the galvanising processes in more detail and the specifications for grades of zinc in commercial use are given in Table 12.

Table 12

Composition of slab zinc ingots required by Standards

Impurity % maximum	UK Specification BS 3436				US Specification ASTM B6−67				
	Zn 1 (99.99)	Zn 2 (99.95)	Zn 3 (99.5)	Zn 4 (98.5)	Special High Grade	High Grade	Inter-mediate	Brass Special	Prime Western
Lead	0.003	0.03	0.35	1.35	0.003	0.07	0.2	0.6	1.6
Cadmium	0.003	0.02	0.15	0.15	0.003	0.03	0.4	0.5	0.5
Tin	0.001	0.001	0.001	0.02	0.001				
Iron	−	0.01	0.3	0.4	0.003	0.02	0.03	0.03	0.05
Total impurities	0.01	0.05	0.05	1.5	0.01	0.1	0.5	1.0	

In the electrolytic process the output is normally of 99.95 per cent purity, and all, or a proportion of, the output can be produced at 99.99 per cent purity or better at some extra cost. To satisfy the market for high grade zinc, the blast furnace and other thermal processes employ a refining stage after metal production. The only process employed at the moment is refluxing, usually the method developed by the New Jersey Zinc Company.

Most blast furnace plants have refluxing units to treat a proportion or all of their zinc output which are capable of producing metal of 99.995 per cent purity.

7.2 THE REFINING OF LIQUID ZINC

Liquid zinc is very much more difficult to refine than lead, copper or tin, because most impurities in these metals can be removed in the liquid state, generally by selective oxidation, but this is not possible with zinc. Some degree of purification can be obtained by liquation, but distillation is of more general application.

Refining by liquation
The liquation method is used to a limited extent to remove excessive amounts of iron and lead and, in some cases, to reduce arsenic levels.

Lead and zinc are somewhat peculiar in that in the liquid state they have only a limited miscibility over a wide temperature range, and this has been used for many years to effect partial purification of horizontal retort metal. From Tables 13 and 14, it will be seen that the solubility of both falls steeply with decreasing temperature. As explained in Chapter 4, the zinc produced in the later stages of the horizontal distillation cycle tends to be high in lead and iron, some of which can be removed by cooling the molten metal under controlled conditions, when the excess tends to separate as the solubility curve is reached.

By cooling a bath of molten metal slowly to 435°C the iron and lead content can be reduced to 0.025 per cent and 1.2—1.0 per cent respectively. The surplus lead sinks to the bottom of the bath and the iron forms a zinc compound (mainly $FeZn_7$) mixed with zinc as a semi-solid, mushy, intermediate layer, between the liquated lead and zinc, called 'hard metal', containing approximately 2 per cent iron.

This principle is used to treat the continuous impure zinc run-off from the bottom of the refluxer lead columns in plants treating metal from the blast furnace.

Using a liquation bath, sometimes designed in the form of a serpentine labyrinth to increase metal retention time and improve cooling temperature control, the lead and iron remaining in the zinc can be reduced to conform to Grade IV specification.

The liquated lead can be removed by air lift pump, and the hard metal by manual or mechanical grab methods, both products being recycled back to the blast furnace. During liquation some arsenic is removed reporting with the hard metal, a typical analysis being:

Zinc	83%
Iron	1.7%
Lead	12%
Arsenic	1.2%

Although an arsenic limit is not quoted in the British Standard Specification for Grade IV zinc, its presence above 0.015 per cent (particularly if associated with traces of indium) will produce a blue discolouration and surface defects in the cast ingots. As both these characteristics are detrimental to the galvanising process, arsenic in Grade IV metal is usually controlled to less than 0.005 per cent, using sodium treatment or aluminium additions to the metal circuit.

Table 13

Mutual solubility of lead and zinc

Temp. (°C)	Pb in Zn	Zn in Pb	Temp. (°C)	Pb in Zn	Zn in Pb
417.8	0.7	2.0	650	9	8
450	1.4	2.3	700	15	12
500	2.3	3	750	24	19
550	4.0	4	775	32	26
600	5.9	6	790	complete miscibility	

Source: Waring, Anderson, Springer and Wilcox, *Transactions of the American Institute of Mining and Metallurgy Engineers,* Vol. 111, pp. 254—263, 1934.

Table 14
Solubility of iron in liquid zinc

Temp. (°C)	Iron %	Temp. (°C)	Iron %
419.4	0.018	600	0.70
425	0.02	700	3.85
450	0.03	800	7.70
475	0.06	(900)	(9.75)
500	0.10		

Source: Truesdale, Wilcox and Rodda, Metals Technology—TP C51-E, American Institute of Mining and Metallurgy Engineers, October, 1935.

Removal of arsenic from liquid zinc

In the presence of sodium or aluminium, arsenic soluble in liquid zinc can be precipitated as an insoluble arsenide, which can be readily oxidised and removed as an arsenate dross from the top of the molten metal. Excess sodium or aluminium may then be removed by further oxidation or by treatment with zinc chloride flux.

Zinc produced from the cooling launders of the blast furnace can contain 0.03—0.06 per cent arsenic, and in the run-off metal from the lead refluxing columns the arsenic content can reach 0.15 per cent. This can be reduced to below 0.003 per cent by sodium or aluminium treatment.

Sodium treatment is normally carried out after the liquation bath process, by first raising the temperature of the zinc to 510°C and batch stirring sodium sticks under the surface of the metal.

The reasons for this are:

— Sodium will act as a de-leading agent as well as a de-arsenicator, and so, for most effective arsenic reduction, the lead content should be circa 1 per cent.
— Some arsenic will have been eliminated in the liquation bath, reporting in the hard metal.
— Sodium treatment ahead of the refluxer columns, without adequate dross separation, can produce excessive dross in the melt baths, resulting in blockages of the feed to the columns.
— The optimum temperature for the sodium treatment is about 500°C+.

7.3 REFINING OF ZINC BY REDISTILLATION

In the early part of this century a number of methods of redistillation using horizontal retorts of various designs were developed in attempts to obtain purer zinc than the horizontal process could produce. These were only partially successful, since a weakness of such methods is that when baths heated below the surface begin to boil, the violent ebullition splashes metal of bath composition on to the walls and roof of the retort. As the zinc evaporates from these drops the lead content increases and consequently its vapour pressure rises, making it difficult to produce refined metal containing less than 0.05 per cent lead. This difficulty was overcome by the development by the New Jersey Zinc Company of a vertical furnace with reflux distillation, which avoided the deficiencies of previous redistillation units. The process produced metal on a large scale, containing over 99.99 per cent zinc.

Today, the metal produced at most thermal plants satisfies the general galvanising market, but to meet the demand for high grade metal a proportion is generally retreated in a refluxing unit of the type originated by the New Jersey Zinc Company. A variation of the New Jersey refluxer was developed by the American Metal Climax Company (AMAX), but although claimed to be a more robust plant, it was less efficient and it has not been used elsewhere.

The New Jersey refluxing unit
The development of the New Jersey refining unit in 1930 was made possible by the availability of silicon carbide (carborundum), which was then a new material, as a refractory. It had outstanding hardness, strength, refractoriness, and, above all, high heat conductivity, making it almost ideal for the construction of the distillation and refluxing columns, which were the basis of the New Jersey design.

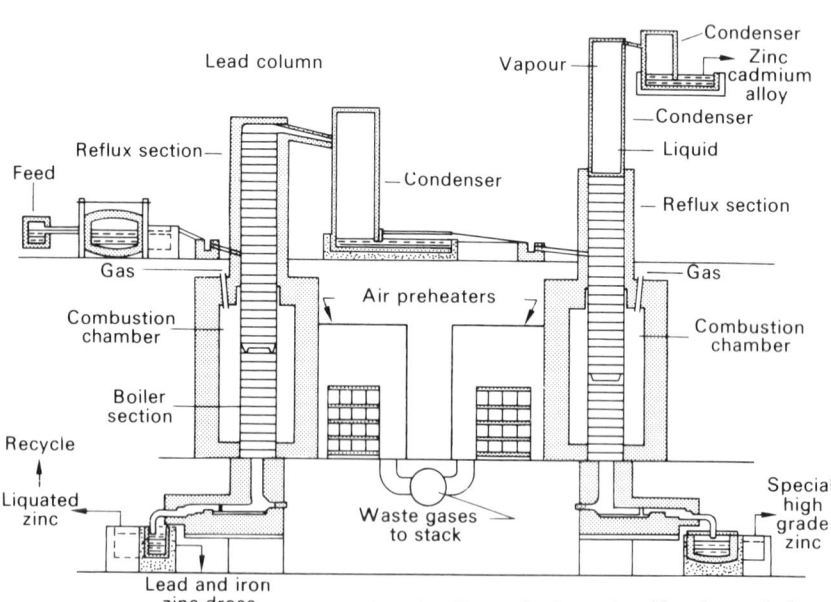

Fig. 34 — Zinc refluxing unit — New Jersey design.

A modern refluxing unit, which is identical in principle with the design of the first units built, is shown in Fig. 34.

For both distillation and refining, columns are built up from carborundum trays, stacked vertically to form a column up to 12 metres high. The trays are rectangular in shape and are normally manufactured in two sizes: standard (48 inches × 24 inches) 1220 mm long × 610 mm wide × 185 mm deep, and large (54 inches × 30 inches) 1370 mm long × 760 mm wide × 185 mm deep. The column can be conveniently divided into two sections called the boiling section and refluxing or upper section.

The boiling section consists of a combustion chamber surrounding the lower part of the column which is built of W trays. These trays have an inverted dished bottom, forming a continuous trough around the inside edge of the tray and holding as much metal as possible against the heated surface, thus achieving maximum heat transfer. The combustion chamber is heated by gas or oil burners. The upper or reflux section of the column is unheated, and the trays are of plain flat bottom design.

In the lead refluxing column, which is the primary boiling column, the reflux section is heavily insulated, using insulation ceramic board backed by insulating refractory brickwork.

The metal to be refined is fed continuously from a holding or melt furnace into the lead column feed tray located above the combustion chamber normally more than half way up the column.

The feed to the column has to be carefully controlled at a very steady uniform rate in order not to upset the delicate balance between heat input, boiling rate and condensation rate, so that violent pressure/suction oscillations within the column are avoided.

Originally, the mechanism for obtaining the very steady feed rate was by mechanically feeding a 1.5–2 tonne feed block (solid feed) into the melt bath by a series of controlled time and depth lowering movements made in response to electrical impulses, to give a steady displacement overflow from a notched weir in the bath outlet to the column feed tube. Developments in refractory technology and availability of new materials has enabled a needle and float valve control system to be used to give more uniform feed rates. In this way liquid metal (liquid feed) can be fed to the melt bath in 3–4 tonne batches direct from the blast furnace circuit, saving the energy required for re-melting the feed blocks in the solid feed system.

As the feed metal cascades down the staggered outlet holes in the bottom of each tray it comes into intimate contact with the metallic vapours rising up the column in a counter-current direction.

Feed rates to lead columns range between 40 and 75 tonnes per day. Approximately 20–30 tonnes per column per day of metal is distilled over into the condenser, depending upon the feed rate and heat input. The excess metal carries out the undistilled iron and lead and other impurities into a supporting sump. The higher feed rates are used when the maximum amount of feed metal has to be de-cadmiumised, and the lower feed rates when maximum production of high grade metal is required. The run-off metal from the lead column is continuously liquated for the removal of iron and lead as previously described. The liquated zinc may be returned to provide additional feed to the columns or it may be cast for sale as Grade IV metal after sodium treatment.

In the unheated reflux section of the column, cooling takes place, which is controlled to encourage some condensation of zinc vapour to provide the reflux action. At the top of the column the vapour, now substantially free from lead, iron and any other high-boiling-point impurities, but containing all the cadmium initially present in the feed metal, passes into a cooling chamber by means of a crossover, to be condensed and collected as a liquid metal in the sump. This metal undergoes another distillation—refluxing process to separate the cadmium from zinc in a second column called a cadmium column.

This column is constructed in the same manner as the lead column, with W trays in the combustion zone and flat bottom trays in the reflux section. The main differences are: a smaller condenser is required, there are generally fewer trays used in the column, and there is little or no insulation round the reflux section because additional cooling is required in this second-stage process.

As the feed metal passes down through the combustion chamber, some 65 per cent of the zinc is vaporised to ensure removal of all the cadmium from the residual metal. The vapour passes up the column into the reflux section where it meets, and is scrubbed by, condensed zinc flowing downwards in a counter-current direction. The vapour leaving the top of the column passes into the smaller condenser where it forms an alloy ranging 10—20 per cent cadmium. The additional cooling for the cadmium column reflux section is required to ensure a heavy reflux action, condensing a high proportion of the zinc vapour, thereby maximising the high grade zinc volume leaving the base of the column, and minimising the volume of alloy produced in the condenser. This alloy is normally cast into a convenient shape, such as a long rectangular strip, to be fed into a much smaller column, tray size 600 mm X 410 mm, for redistillation to produce cadmium metal containing approximately 0.5 per cent zinc, which is removed by a caustic soda/sodium nitrate treatment to produce high purity, saleable grade cadmium.

The capacity of the small column is 3—4 tonnes feed per day and up to 300—350 tonnes per annum cadmium. At some plants the cadmium column vapour is shock-chilled in a cannister, collected as a dust from which the cadmium can be recovered.

At the bottom of the cadmium column the refined zinc is collected in a supporting sump and laundered to the casting area to produce saleable product shapes. Great care must be taken so that this metal does not make contact with any metal, iron, support steelwork, tools, etc., so that there is no contamination. The quality of metal produced generally exceeds 99.995 per cent zinc (special high grade), containing 0.001–0.003 per cent lead, 0.0005–0.001 per cent iron and 0.001–0.002 per cent cadmium.

The lead boiling columns of the type described have a boiling capacity of approximately 30 tonnes per day. The feed to such columns ranges from 45–75 tonnes per day, which may include 20–30 tonnes of recycle metal. Two lead columns are generally linked to one cadmium column so that the daily capacity of such a unit is of the order of 60 tonnes of high grade zinc and 1–2 tonnes of alloy, and the balance, being the lead column run-off, is available for casting as Grade IV metal or recycling.

The hot waste gas (1050–1100°C) leaving the combustion chamber passes through a recuperator and heats the air required for combustion to 750–800°C. The temperature of the lead column combustion chamber is maintained at 1200–1300°C and the cadmium column is 1150–1220°C.

The amount of heat required to produce 1 tonne of purified zinc is approximately 6 GJ with good practice, but the heat required to vaporise 1 tonne of zinc from cold metal is 2.27 GJ per tonne: hence the process is thermally inefficient.

It is appropriate here to mention some operating variations from the standard unit of two lead columns and one cadmium column.

The high grade capacity of a refluxer plant can be increased by partial or total recycling of the lead columns' run-off metal. As this metal is already de-cadmiumised a total recycle feed to a lead column will produce a condensate containing very little cadmium, which generally meets a high grade or alloy grade specification without the requirement to process through the cadmium column. Such a column handling only recycle feed is referred to as a reboiler, or reboil column, and if specifically built for this purpose generally contains fewer refluxer section trays. Some refluxing action is required, however, to control the lead levels in the condensate metal in order to meet the normal high grade limitation of 0.003 per cent lead. If an intermediary or alloying grade metal is required and higher levels of lead can be tolerated, a short reboiling column can be used. Such columns have virtually no refluxer section and the crossover to the condenser is located just above the feed tray.

It must be noted that the run-off metal from reboiler columns, having gone through a second distillation process, will contain a higher concentration of impurities such as lead and iron, if not previously liquated, and also higher concentrations of tin, copper and some minor elements which are virtually unaffected by liquating. It is the level of this latter group of impurities which determines the quantity of metal which can be recycled to augment the lead column feed and still enable the run-off to conform to the saleable Grade IV specification after the liquation process.

In plants using a total recycle system, the bleed products from the liquation process can be used as a first-stage concentration for further processing of any desirable elements, some examples being germanium, tin and lead, etc.

Principles underlying the process

The removal of impurities, mainly iron, lead and cadmium, can be accomplished because at the boiling point of zinc (907°C), cadmium is well above its own boiling point (767°C); the vapour pressure of lead is low; and iron, aluminium and other impurities are virtually non-volatile. Therefore, in the first stage, carried out in the lead column, cadmium is evaporated almost completely and leaves, with the volatilised zinc, to be separated in the cadmium column. Iron, aluminium and other non-volatile impurities are carried down by the unvolatilised zinc and pass out from the bottom of the column.

The position of lead is more complicated. Lumsden has shown that, in equilibrium with a dilute solution of lead in zinc, at the boiling point of zinc the lead/zinc ratio in the gas is only 0.005 times that in the liquid [2] and thus only a little reflux is theoretically necessary to effect the required degree of separation. This is as well, since the use of any bubbling system is precluded, owing to the high density of zinc, and high pressures in the columns cannot be tolerated. The contact between liquid and vapour is poor and there is little scrubbing action.

Assuming the average lead content of the metal in the column is 2 per cent, that of the vapour in equilibrium with it will be approximately 0.01 per cent. The vapours entering the upper unheated section will therefore contain this amount of lead, plus an unknown amount of liquid lead carried out as spray arising from the high velocity of the vapour leaving the boiling metal. This spray is trapped by the tortuous path of the gas in the upper section and is removed without difficulty. The lead present as vapour can be removed by the condensation of a small proportion of the zinc. At each stage the condensate contains 200 times the lead/zinc ratio in the vapour at that stage, so that the condensation of only 2–3 per cent of the zinc would reduce the lead content to 0.003 per cent or less. Thus the amount of condensation in the upper section required to effect the necessary degree of lead purification is very small.

The cadmium content of the metal entering the cadmium column is normally of the order 0.1–0.4 per cent. Concentrations up to 0.5–0.7 per cent have been known, and under these circumstances it is very difficult to lower the cadmium level to $\leqslant 0.002$ per cent even by considerably reducing the volume of feed and increasing the column firing rate. Since the boiling point of cadmium is considerably below that of zinc, in the normal range of concentration it can be removed almost completely in the boiling section of the column by evaporation of approximately 65 per cent of the zinc feed.

Almost all of this zinc boil-off is recondensed in the upper unheated section of the column, and runs back to join the feed metal and to pass through the boiling section repeating the cycle, purifying the zinc which reports at the bottom of the column. The requirement to recondense almost all of the zinc in the reflux section of the cadmium column illustrates the difference in mode of operation of lead and cadmium columns. In order to facilitate this condensation, additional surface area is provided in some cadmium columns by inserting free-standing carborundum pieces in the bottom third or half of the reflux trays. The cadmium vapour leaves the top of the column and is condensed to form an alloy containing 10–20 per cent cadmium for further refining as previously described.

Casting techniques

As a result of pressure from safety considerations and helped by the availability of improved refractories, a considerable advance in casting technology took place during the late 1970s.

The standard production shape is a 25 kg rectangular plate, normally carrying the producer's logo and grade of zinc imprinted on the bottom surface, packaged into 1 tonne pallets containing 40 plates. The plates are cast on a carousel wheel or, more commonly nowadays, on a straight line, continuous mould, chain casting machine. Large blocks are generally of similar dimensions from different producers and range from 0.8 tonnes to 1.0 tonnes to larger or more specialised shapes. The presence of cooling cavities, particularly in the larger shapes, allowing condensation or rainwater seepage to enter the cavity through any surface cracks, has led in the past to occasional explaosions when the blocks are charged directly into a molten bath of metal, as is customary in the hot-dip galvanising lines. In order to avoid cavity formation, and eliminate these accidents, a system of water cooling and top heating has been developed.

On the straight line casting machine, the bottom of the moulds are cooled using a curtain of water from a series of sprays, with a collection trough and splash guard plates so designed to prevent water droplets escaping and reaching the surface of the metal. The top surface is heated, using strategically located gas burners, to prevent the premature solidification of the surface and so to allow the solidification to take place from the bottom upwards. The burners, sprays and casting rates are co-ordinated to ensure that each plate is completely solidified before being discharged from the machine.

For the larger blocks, the moulds are mounted in individual water-cooled jackets, supplied by a continuous flow of cold water entering at the bottom and overflowing from the top of the jacket. The surface of the block is top heated to ensure the unidirectional solidification of the block from the bottom upwards. This heating is adjusted and controlled to ensure the molten liquid pool on the top surface is the last to solidify and give a good surface appearance to the block.

Large blocks are normally removed from the mould using an overhead crane to lift wedges (or dogs), inserted on the sides of the moulds, which are then knocked out manually or by mechanical means. For this reason top heating units consisting of gas heated radiation mantles or electric heating bars are often mounted on a wheeled carriage with track, to allow the unit to be conveniently moved over or away from the block for crane access.

Improvements in refractory materials have enabled cheap and efficient pumps to be designed, driven by air motors, for use in lifting and delivering molten zinc at controlled temperatures, volume and flow rates, to casting machines or block moulds, thus ensuring uniformity of product weight, appearance and quality control. The top heated–bottom cooled, straight line casting machine delivers a plate on to a synchronised moving chain conveyor passing through a water-cooling trough, on to an automatic stacking machine. The 1 tonne pallet so produced is compressed and straightened, weighed and identified, strapped automatically and presented ready for delivery to the customer.

Good insulating materials with 'non-wetting' characteristics have enabled molten zinc to be laundered over long distances with no great loss in metal temperature. This has eliminated the need for external heating which, as well as being expensive, caused the formation of zinc oxide, requiring manual attention for cleaning and recycling of the dross back into the system.

These factors have significantly contributed to reducing unit costs and influenced the design layout and automation of the casting and handling of zinc in a modern casting shop.

The refluxer operation is expensive, and places thermal processes at a disadvantage compared with the electrolytic method which can produce, as required, zinc of 99.99 per cent purity, or better, at little extra cost. However, in order

Fig. 35 – Zinc melting and casting plant at the Risdon, Tasmania, works of the Electrolytic Zinc Company of Australasia Ltd.

to make valid comparisons it is necessary to evaluate the total processes, where the advantages of the sinter, blast furnace and refluxer plants in tolerating a cheaper and wider range of raw materials with resultant by-products must be compared with the roasting, leaching, purification and electrolytic process with high electric power requirements. For the refluxer process the cost of fuel is relatively high, as is that of maintenance, although in recent years column life has been improved and extended with campaigns over three years being achieved. Liaison between the user and manufacturer to improve tray design, firing and quality; conversion to cleaner and more efficient natural gas firing; introduction of liquid feed systems; use of a mix of large and standard tray columns; reboiler columns to suit individual refinery needs — all these factors have contributed to reducing unit operating costs. Another major cost even with good operation and use of fluxing agents is that some 1.5–2.0 per cent of input zinc is oxidised during the process, and although this is not lost, it has to be recycled back to furnace or sinter circuits. The process, however, is an elegant application of distillation and refluxing techniques at high temperatures and with liquid metal. It reflects credit on those who developed it, and at the moment it remains the only commercial method for refining low grade zinc to meet Grade I specification. A description of the refluxing operation at Cockle Creek is given by Tozer and Cunningham [3].

REFERENCES

[1] Tainton, U. C., Taylor, A. C. and Ehrlinger, H. P., Lead alloys for anodes in electrolytic production of zinc, *Transactions of the American Institute of mining and Metallurgical Engineers,* February Meeting, 1929.

[2] Lumsden, J., Thermodynamics of lead–zinc alloys, *Discussions of the Faraday Society,* No. 4, pp. 60–68, 1948.

[3] Tozer, B. W. and Cunningham, D. A., The zinc refluxer at Cockle Creek, *Proceedings of the Australasian Institute of Mining and Metallurgy,* 233, March, pp. 53–59, 1970.

8

Recovery of zinc from scrap and low grade materials

The recovery of zinc from scrap and low grade materials will become increasingly important as the necessity for conservation of the metal becomes more apparent.

8.1 SOURCES OF SCRAP

The main consumption of primary zinc is in galvanising and die casting, and in the rolling and brass industries. During manufacturing and fabricating operations in these industries, a proportion of scrap is made, and some of the metal is oxidised or degraded. As much as possible of the metallic zinc in these secondary materials is recovered and re-used on site, but a proportion must be recovered elsewhere. This proportion is highest in the galvanising industry where approximately 25 per cent of the virgin metal input is degraded to dross or ash, which, although the metallic zinc content is high, is so contaminated with oxide, iron or chlorine that it cannot be re-used in the galvanising baths, but must be treated outside this industry.

Much of this residual material is used for the production of zinc oxide, zinc dust and chemicals, but in the United Kingdom the demand in these industries exceeds the indigenous supply and some imported material and virgin zinc is used. A proportion of the residuals is too oxidised or contaminated for treatment other than in a zinc—lead blast furnace. The final copper-containing residues from the brass industry are also treated this way and then approximately 80 per cent of the copper can be recovered, as well as the zinc present.

Should the production of secondary drosses and scrap material exceed the capacity of the ancillary industries to absorb them, they must be treated by the blast furnace, since the electrolytic process would have difficulty in coping with the amounts of chlorine and aluminium they are likely to contain.

A further and increasing supply of zinc-containing residues arises from scrapped automobiles and other similar equipment. The recovered brass from such operations is reasonably high but, because of its lower value, much of the zinc scrap consisting mainly of die castings is still lost. Most of the recovered scrap is used for zinc dust and oxide production.

Other possible sources of secondary zinc which are not utilised in the UK at present are the flue dusts arising in the steel industry. Many iron ores contain small quantities of zinc, which is driven off in the iron blast furnace to collect in the flue dusts. Whilst much of this dust is recycled, a build-up of zinc occurs until is must be discarded. Zinc also collects in the flue dusts arising in both oxygen and electric furnace steel-making operations, largely from the treatment of galvanised steel scrap. Whilst the zinc content of such dusts is low (1.6 per cent) they cannot continue to be discarded, and methods of treatment to recover both zinc and iron are being developed.

8.2 RECOVERY FROM LOW GRADE ORES

The recovery of zinc from low grade ores presents problems. All the existing primary methods of zinc production require high grade charge, and their production costs rise rapidly as the zinc content of the ore treated drops, although the zinc–lead blast furnace process is somewhat less sensitive in this respect than its competitors.

A proportion of the world's zinc reserves is locked up in oxidised ores which contain 10 per cent or less of zinc and which cannot be concentrated by existing mineral dressing methods. There are also tailings and residual material from primary processes containing appreciable quantities of zinc and lead which it is wasteful to discard. For this type of material the Waelz process is at present the only method of concentration and its importance is likely to grow. For the future upgrading of these ores and residuals, the industry is in need of a new method, less demanding in fuel and other operating costs.

The Waelz process

The Waelz process was developed in Germany by Krupp, and the first commercial installation was installed at Luenan in Upper Silesia in 1925. The plant consists essentially of a long rotating kiln, refractory lined and fired internally by oil, gas or pulverised coal. Most kilns in use are approximately 70 m long and 4.5 m diameter, but at Miasteczko in Poland eight kilns 95 m long and 4.2 m diameter are installed, ranking as the world's largest Waelz installation. The speed of rotation can be varied, and lies generally in the range 0.6–1.2 revolutions per minute, depending on the flow characteristics of the charge. The kiln is inclined, the angle of inclination varying usually from $2°$ to $5°$.

The ore or residue is crushed to approximately 10 mm, care being taken to avoid excessive production of fines, and is then mixed with coke breeze (0.5–6 mm) until the carbon content is 20–25 per cent of the ore treated. The mix is fed into the top end of the kiln and passes slowly along its length under the influence of the rotation and inclination. Retention time in the kiln is usually 8 hours.

Heat is supplied by a burner at the discharge end, and the temperature is raised to 1000–1200°C in the hottest zone. The temperature is controlled, and must not be allowed to exceed the softening point of the charge. The heat required to maintain this temperature is approximately 1.55 GJ (equivalent to 50 kg of coke) per tonne of ore treated. Under these conditions, owing to the presence of carbon, the atmosphere in the bed contains sufficient carbon monoxide to reduce the oxides of zinc and lead. These metals are volatilised and then reoxidised in the atmosphere above the bed either by oxygen or carbon dioxide, reforming oxides as fume, which is swept out of the kiln by the gas flow.

The dust- and fume-laden gases are drawn first through chambers to precipitate dust and then through a system of cooling flues, where the temperature is dropped to 120°C. After passing through a fan providing the suction for the system, they enter bag filters where the oxides are collected in terylene or glass fibre bags. Electrostatic precipitators are also used, but to a lesser extent.

At some plants the fume is treated directly, either by electrolysis, or in a zinc–lead blast furnace. In a number of cases where electrolysis for recovery of zinc is to follow, an additional calcination stage is used with the advantage of removing harmful impurities such as chlorine or fluorine. Calcination also densifies the oxides which facilitates sintering if fed to a blast furnace.

The hot spent charge leaves the kiln through a chute, and is usually quenched in water for convenience of handling. Partial slagging of the charge occurs in the kiln, and if this is excessive, limestone may be added to reduce fluidity. With most charges, accretions tend to build up on the kiln lining, particularly towards the firing end. They can be held in check by barring periodically and sometimes they can be removed by melting, but if this is not successful the kiln has to be emptied and the offending accretions removed when cold. Attack on the lining also occurs, which can be limited by water cooling, but again this tends to reduce campaign life, and is used only in an emergency.

Under normal conditions, with charges which do not fuse too readily, over 90 per cent of the zinc and lead can be eliminated, and the zinc and lead content of the spent charge reduced to 2 per cent and 0.2 per cent respectively. Iron and copper are not volatilised and remain in the residues. Silver is removed only partially to the extent of 35–40 per cent. Any germanium present is volatilised as the dioxide, and if worthwhile can be recovered from the fume. Approximately 20 per cent of the carbon is unburnt and remains in the residues. With some charges, such as willemites, which tend to fuse readily, excess carbon is added, sometimes up to 50 per cent of the charge weight. This reduces the strength of the accretions, which are then less prone to remain attached to the kiln lining. A proportion of the additional carbon can be recovered by jigging the residues, and returning it to the kiln.

Within broad limits a kiln can treat a certain tonnage of ore per day and therefore its production of oxides is dependent directly on the zinc and lead content of the feed. For materials with a zinc content lying in the range 10–14 per cent, capacity can be expressed as 0.7–1.0 tonne ore per cubic metre of kiln volume – the variation depending mainly on ore sizing and fusibility [1]. For higher grade ore (16–20 per cent zinc) kiln capacity can be better expressed in terms of metal evaporated, and a factor of 125–140 kg of zinc plus lead per cubic metre of kiln volume can be taken.

The process is thermally inefficient. Although the heat required to reduce the zinc and lead oxides in the bed is liberated again, as reoxidation takes place in the kiln atmosphere, the heat loss from the shell is high, and much is also lost in the gases which leave at temperatures of 500–600°C, and in the spent charge which is discharged at 900–1000°C. As a result the coke requirement of the process is considerable, and can reach 30 per cent of the weight of charge treated, of which 25 per cent is added as reducing material to the charge and 5 per cent is the coke equivalent of the oil or gas required by the burner.

As has been stated, the economics of operation depend entirely on the grade of feed treated, since the coke and fuel consumption and most other costs are largely fixed. With relatively high grade material (zinc and lead equal to 25 per cent) approximately 1.4 tonnes of coke (reducing coke plus burner fuel coke equivalent) is required per tonne of metals recovered. With low grade charges (zinc and lead equal to 10 to 12 per cent) the coke required can be as high as 6 tonnes per tonne of metals recovered, and, although the coke used can be relatively low grade and in breeze form, its cost is still a heavy charge on the process.

In spite of its low thermal efficiency the process performs a useful and varied role in recovering lead and zinc from low grade ores and residues. At the Porto Vesme plant in Sardinia, 500 tonnes per day of ore containing 15 per cent zinc and 2 per cent lead is concentrated in two Waelz kilns to produce oxide for treatment in a zinc–lead blast furnace. The Miasteczko plant in Poland is of particular interest since approximately 3500 tonnes of ore containing only 6.8 per cent zinc and 0.9 per cent lead are treated in eight large kilns to produce fume for similar treatment. At Aizu in Japan 100 tonnes per day of electrolytic plant residues containing 22 per cent zinc and 3.4 per cent lead are treated in two relatively small kilns. A similar operation is carried out in Zambia where residues from an electrolytic plant and tailings from the concentrator are treated with other zinc containing materials in a kiln. The zinc oxide–lead oxide fume is incorporated in the feed to the zinc–lead blast furnace at the site. With characteristic thoroughness the Japanese treat the kiln residues to recover gold and silver and also gallium and indium [2]. Descriptions of other Waelz operations are given in other references [3, 4].

In Japan, Germany, and France, steel plant dusts are treated by kilning, and the zinc oxide–lead oxide fume is incorporated in the feed to a zinc–lead blast furnace, either at the sintering stage, or directly to the furnace in the form of hot-pressed briquettes.

In the hot-briquetting process, the oxides are heated in a rotary furnace, and are then briquetted under pressure, in a roll press, at a temperature of 450°C–600°C. At these temperatures, a certain amount of plastic flow occurs under pressure, together with recrystallisation, and a strong bond between the particles develops. This maintains its strength at higher temperatures. Thus the briquettes, when fed to a blast furnace, retain their shape with the minimum of decrepitation.

8.3 SLAG FUMING

In most ore deposits zinc and lead occur together and, despite the high efficiency of modern flotation techniques, zinc concentrates generally contain some lead, and most lead concentrates contain zinc. In orthodox lead blast furnace smelting, a proportion of the zinc is volatilised and caught as zinc oxide in the fume but most is retained in the slag tapped from the furnace bottom; two factors contribute to this.

In lead smelting there is insufficient carbon in the charge to produce the high carbon monoxide–carbon dioxide ratio necessary to reduce zinc oxide in quantity, and the relatively low lime–silica ratio (0.5–0.7) in the slags produced, lowers the reactivity of zinc oxide dissolved in them, thus lessening the tendency for it to be reduced and then volatilised. This is the opposite of the conditions in the zinc blast furnace where maximum reduction and volatilisation of zinc is the main aim. This is achieved by running with much stronger reducing conditions, owing to the higher proportions of coke in the furnace burden, and from the use of high blast air preheat temperatures (temperatures of 1050°C have been used). The slags also have a higher melting point and thus higher temperatures can be reached in the slagging zone. At these temperatures the activity of zinc oxide is increased, and it is therefore more easily reduced.

Table 15 shows analyses of lead blast furnace slags which are typical of modern practice. It will be seen that the slags carry considerable quantities of

Table 15
Lead blast furnace slag analyses

	Herculaneum, Missouri, US	Buick, Missouri, US	Glover, Missouri, US	Trail, British Columbia, Canada	Port Pirie, Australia
SiO_2	20	22.9	23.5	20.8	21.0
CaO	9	18	12.1	10.3	14.7
FeO	33	31	32	35	25.6
Zn	15	11.5	10.8	17.5	18.5
Pb	3.5	3.5	3.8	2.5	2.3

Source: American Institute of Mining and Metallurgical Engineers World Symposium on Mining and Metallurgy of Lead and Zinc, 1970.

zinc and lead, a loss which is appreciable unless recovered by subsequent treatment. A method widely used for this is slag fuming, and since the first commercial slag fuming unit was installed at East Helena in 1927 by the Anaconda Copper Mining Company, units have been built at a number of other locations.

In this process molten slag direct from the lead blast furnace, together with a proportion of cold granulated slag if it is necessary to treat dump material, is fed as a batch into a furnace which is a steel tank formed of water-jacketed sections. In a typical installation the internal dimensions of the tank are 6.5 m by 2.5 m by 3 m and such a furnace could treat a batch of approximately 40 tonnes of slag containing 10–15 per cent zinc. Along each side of the furnace are fitted twenty-one 50 mm diameter tuyeres through which pulverised coal and air are injected below slag level, the proportion of coal to air being carefully controlled. During charging, the air is in slight excess so as to generate the maximum amount of heat to melt all the slag and to raise the bath temperature to 1200°C. After this stage the coal-to-air ratio is increased so as to generate the appropriate carbon monoxide–carbon dioxide ratio to enable reduction of zinc oxide to take place. At this stage the evolution of zinc and lead is copious, and the zinc oxide content of the bath drops at a roughly linear rate [5]. After approximately two hours the zinc oxide content has been reduced to 2.5–3.0 per cent and the lead content below 0.5 per cent. At this point the coal-to-air ratio is reduced to raise the temperature of the spent slag prior to tapping.

Approximately 25–30 per cent of the charge weight of good quality coal (27.6 MJ/kg) is required, depending on the proportion of cold slag treated. This is equivalent to 1.5–1.7 kg of coal per kg of zinc plus lead in the fume. From 5.5 to 7 tonnes of steam are recovered per tonne of coal burnt, in waste heat boilers.

There are differences of opinion as to the detailed mechanism of reduction and the rate of limiting factors in the slag fuming bath. Kellog has claimed [6] that owing to the high turbulence created by the air in the bath, the area of gas–slag interface is very high (above 900 m²) and the rates of elimination obtained seem lower than would be expected. In a Bulgarian paper [7] it is stated that the limiting step in the process is the diffusion of zinc in the molten layer at the gas–slag interface. At Trail, British Columbia, an analysis was made of the fuel requirements of the slag fuming bath based on the assumption that slag, air and fuel reach equilibrium in the furnace [8]. Quarm claims that this assumption is unjustified, and suggests that the rate of reduction of zinc oxide in the slag is mainly controlled by a pseudo-first-order reaction involving ferrous oxide [9].

The main reducing mechanism certainly involves carbon and carbon monoxide. The hydrogen in the volatile matter of the coal appears to play a minor part. For some time it was claimed that oil or natural gas could not be used, but at Plovdiv in Bulgaria on a furnace treating 200 tonnes per day of slag, a successful conversion from pulverised coke firing to heavy oil has been claimed.

It is reported that the oil consumption is equivalent to 1.65 tonnes per tonne of zinc removed from the input slag.

No successful attempt has yet been made to improve the heat economy in a fuming furnace by enriching the blast with oxygen. At a slag fuming plant built at Port Pirie in Australia, metal recuperators are used after the waste-heat boilers, to preheat the air required for the furnace to 600°C. As a result, the coal consumption per tonne of zinc eliminated has been reduced to 1.02 tonnes. This represents a considerable saving.

A number of slag fuming plants are in operation throughout the world as an adjunct to lead blast furnace smelting. The process performs a useful role in recovering zinc, as well as lead, which would otherwise be lost, but in the form of oxides which require further treatment. Devotees of the zinc–lead blast furnace might argue that the zinc oxide containing slags should not have formed in the first place.

REFERENCES

[1] Friedrich, C., Heinrich, W. and Gornitz, M., Optimisation of Waelz process, *Neue Hutte,* Vol. 16, no. 8, pp. 457–461, 1971.

[2] *American Institute of Mining and Metallurgical Engineers World Symposium on Mining and Metallurgy of Lead and Zinc,* Vol. 2, p. 412, 1970.

[3] Morrison, C. W., The Waelz process, *Zinc,* ed. C. H. Mathewson, New York, Reinhold, 1959.

[4] Cross, H. E. and Read, F. O., Waelz treating of complex zinc lead ore, *American Institute of Mining and Metallurgical Engineers World Symposium on Mining and Metallurgy of Lead and Zinc,* Vol. 2, p. 918, 1970.

[5] Pas, S. K. and Sarkar, S., Recent developments in slag fuming process, *Tech. Symposium National Metallurgical Laboratory Jamshedpuir, India,* 4–7 Dec., 1968.

[6] Kellogg, H. H., Computer model of the slag fuming process for recovery of zinc oxide, *Trans. of the Metallurgical Society of the American Institute of Mining and Metallurgical Engineers,* Vol. 239, pp. 1439–1446, Sept. 1967.

[7] *Chemical Abstracts,* Vol. 81, 94368Y, 1974.

[8] Bell, R. C., Turner, G. H. and Peters, E., Fuming of zinc from lead blast furnace slag, *Journal of Metals,* March, pp. 472–477, 1965.

[9] Quarm, T. A. A., Slag fuming – kinetic or thermodynamic, *Engineering and Mining Journal,* Vol. 169, Jan., pp. 92–93, 1968.

9

Pollution control in the zinc industry

As stated in the section on 'The Biological Significance of Zinc' (Chapter 14), zinc can generally be regarded as non-toxic; in fact its presence in small quantities seems to be essential for many forms of life. Unfortunately for those concerned with its extraction, zinc in nature is almost always found in the sulphide form, associated with lead and cadmium, and almost all of the flotation concentrates providing the raw material for the smelting industry contain small but significant quantities of these elements. Both lead and cadmium are very toxic and can be tolerated only in extremely small concentrations. The environmental problems which press heavily on the zinc smelting industry arise almost entirely from their presence. There are other hazards such as those from handling arsenic, from acid spray arising in electrolytic plant tank houses, and from the control of sulphur dioxide emissions from roasting and acid plants, but these are minor compared with those arising from the presence of lead and cadmium.

9.1 TOXICOLOGY OF LEAD AND CADMIUM

The toxicology of lead and cadmium is now well established. Even up to the early days of this century cases of lead poisoning were only too prevalent, where the whole body was damaged – especially the nervous system, the gastro-intestinal tract, and the blood-forming tissues. Such cases were caused by relatively heavy absorption of the heavy metal, which rarely occurs today. At the present time, as far as adults are concerned, the debate relates to the levels of absorption at which relatively minor symptoms such as reduction in nerve velocity are observed. Of recent times very few cases have been documented where non-occupationally exposed adults have shown symptoms which can be attributed definitely to excessive lead absorption. In the case of children, serious concern continues that very low levels of lead absorption may affect the development of the brain and central nervous system, retarding intellectual development and reducing the level of eventual attainment. The tests on which these results are based are complex, giving results which in many cases appear only marginally conclusive. However a recent Royal Commission considered that the evidence was sufficient to justify a recommendation that additions of lead to petrol should be banned.

Exposure to cadmium and its compounds leads to accumulation in liver and kidney, interfering with the vital functioning of these organs. Like lead, its natural rate of elimination from the body is low, and it is therefore a cumulative poison. However, lead tends to accumulate in bone and not in liver or kidneys, and its long-term effect in the case of adults is less serious. Cadmium fume is extremely hazardous and lengthy exposure to it can lead to lung damage in the form of emphysema.

Some measure of the relative toxicity of zinc, lead and cadmium can be seen in the World Health Organisation's proposals for the upper limits of impurities in drinking water as shown in Table 16.

Table 16

Metals in drinking water: recommended upper limits of the
World Health Organisation, 1971

Metal concentration (mg/litre)		
Zn	Pb	Cd
5	0.10	0.01

9.2 ENVIRONMENTAL STANDARDS

Owing to the growth in recent years in public concern over the effect of environmental pollution, in all industrial countries a flood of legislation has appeared or is proposed, imposing severe limits on both gaseous and liquid effluents released from metallurgical and other plants. There is an unfortunate lack of consistency both in the philosophies adopted and in the limits proposed for pollution control. The confused situation at the present time is well described by A. K. Barbour [1].

The protection of our environment is a matter of vital concern to all of us, and the public interest in the problem is to be welcomed, but it has brought many difficulties to the zinc industry. These can be considered under the following headings:

1. Effect on areas outside the plant

    (a) from emissions to atmosphere
    (b) from liquid effluents
    (c) from solid waste disposal

2. Effect on the workforce within the plant

Environmental standards on atmospheric emission

As has been stated, with zinc blast furnace plants (and also to a lesser extent with electrolytic plants), the main environmental concern arises from the presence of lead. In the United Kingdom all atmospheric pollution is regulated through the Industrial Air Pollution Inspectorate, which was formed over 100 years ago and is now part of the Health & Safety Executive. On the whole, over the years, it has won the respect of industry, and in general there has been useful co-operation between the two. Until recently, in the UK no standards for ambient lead in air had been laid down, emissions and chimney heights being regulated on an individual basis using the 'Best Practical Means' philosophy, which implies that each environmental problem had to be tackled individually, and solved using the best technology and equipment available. To give teeth to this philosophy, the Inspectorate has laid down limits for total quantities of lead to be emitted into the atmosphere. For a plant with a gaseous emission of 4000 m^3/min the following limits must not be exceeded:

	Individual stack concentrations $\mu g/m^3$	Total mass emission kg/h
Lead as metal	0.0115	5.4

There are also statutory commitments for plants emitting lead to monitor stack fall-out on surrounding ground and vegetation.

The UK is now being forced to relinquish its adherence to the 'Best Practical Means' philosophy. Most countries now impose ambient air standards. In Europe a value of 2 $\mu g/m^3$ (annual average) for lead tends to be applied, and under EEC directives this is to be adopted in the UK. This is a stringent requirement, but at most zinc blast furnace plants and lead smelters it is considered attainable. In the United States, the Environmental Protection Agency has announced an even stricter standard of 1.5 $\mu g/m^3$. Whether this can be attained by lead smelters and what medical evidence is available to support its imposition is not clear.

Sulphur dioxide

In the UK, current Alkali Inspectorate requirements for emissions from acid plants treating sulphur dioxide arising from zinc blende roasting demand that the sulphur lost to the air as acid gases shall not be greater than 0.5 per cent of the sulphur burned or, in the case of sinter-based plants, of the sulphur dioxide fed to the acid plants. The waste gases must be substantially free from acid mist. Within the EEC there is a regulatory limit of 350 $\mu g/m^3$, which the sulphur dioxide content at ground level in the area of maximum impingement must not exceed.

With modern double conversion/double absorber acid plants, it is possible to operate well within these limits, and to attain the standards governing sulphur dioxide emissions is generally much less onerous than to attain that governing lead.

The position may change if the present concern about the long range environmental damage under the general heading of 'Acid Rain' results in demands for a reduction in the amount of sulphur dioxide emitted by industry. A number of European countries (but not yet the UK) have already agrred to take immediate steps to reduce all sulphur dioxide emissions by 30 per cent [2]. This could be done by the zinc smelting industry in this country, at some extra cost, by inserting an additional scrubbing system. The gaseous effluent of the zinc smelter at Avonmouth, producing 90,000 tonnes per annum of zinc and 35,000 tonnes of lead, puts some 600 tonnes of sulphur dioxide into the atmosphere annually. The Central Electricity Board is a much more serious offender, since the waste gases from its coal fired power stations contain over 1 million tons of sulphur dioxide, and this at the moment is emitted untreated.

Cost of pollution control in the zinc industry
It is not possible to specify accurately the considerable cost of pollution control to the industry since so much depends upon local conditions and on the standards which must be reached. Atkins and Lowe, investigating the problem for the Department of the Environment in 1977, studied the zinc plant at Avonmouth. They estimated that the cost of pollution control at this plant was equivalent to between £10 and £20 per tonne of zinc produced [3].

Liquid effluents
As in the case of gaseous effluents, the philosophy in the UK on the control of liquid discharges tends to differ from that of other countries. In the UK, 'consents' or licenses to discharge must be obtained from the local Regional Water Authority, and the limits not to be exceeded are related to the uses to which the systems receiving the discharges are put. These are classified in various ways, such as potable, estuarial, recreational, sewerage, etc., and the limits imposed in the 'consents' are correspondingly varied. In most other countries the imposition of fixed standards independent of the destination of the effluent is generally favoured, which sometimes forces precautions to be taken which under the UK system would be unnecessary.

The required degree of purification from heavy metals can generally be obtained by neutralising with lime to pH 9.5–10.0, in a single-stage precipitator, with adequate settling facilities. Under correct conditions, removal of zinc, lead and cadmium, both in solution and as suspended solids, to below a combined total of 5 mg/l can be obtained without great difficulty, and this will satisfy most present standards. Cyanides can be removed by aeration. Almost complete removal of arsenic can be achieved by addition of ferrous sulphate. All precipitated sludges are recovered and returned to the roasting operation.

The storage of residues from zinc plants can cause difficulty if they contain soluble salts. This is so in the case of jarosite or goethite precipitates from electrolytic plants, which can contain up to 1 per cent of soluble zinc. Such residues must be stored in dumps with impervious foundations so that all liquors draining from the dump can be collected and returned to the plant for purification. This is a nuisance since the precipitates are bulky and in a fine state of division. The problem is greater with jarosite precipitates (iron content 25–28 per cent) than with goethite (iron content 40–45 per cent) since the bulk to be stored is appreciably less in the latter case.

In this respect the blast furnace process holds a definite advantage over the electrolytic method, since most of the residues produced are contained in the discard slag. Although this contains heavy metals such as zinc, lead, arsenic and copper these are in insoluble form and the slag can be stored outdoors, without risk of surface water contamination. In some cases it can act as a filling material and has some value.

In-plant hygiene
The health of the operatives is a prime consideration in any manufacturing process, and this is especially so in the case of the zinc smelting industry, where lead, cadmium and other toxic metals are always present. Although problems arise in electrolytic plants owing to the presence of these toxic elements, the blast furnace process is regarded as lead smelting and the problems of in-plant hygiene, being much greater, can dominate the operation.

Adequate ventilation at all points where emissions to the atmosphere can arise is the first consideration, and the skill of the ventilating engineer is in great demand. The large volumes of ventilating air required must all be filtered before final discharge, as must all emission to stacks. Some measure of the care required can be gathered from the fact that at Avonmouth, where the blast furnace produces about 90,000 tons of zinc and 35,000 tons of lead per year, the total amount of lead from all plant gaseous emissions must not exceed 5.4 kg per hour.

Monitoring of the in-plant atmosphere must be carried out continuously using static sampling equipment. Personal samplers attached to operators considered to be at risk must also be used. The Threshold Limit Value (that concentration in atmsophere which can be tolerated by an average person exposed for eight hours per day for a 40-hour week) in the case of lead has been fixed in the UK at 150 μg/m^3. The corresponding value in the case of cadmium is 5 μg/m^3. It is difficult but not impossible to keep all sections of a blast furnace plant below these values. Good washroom facilities must obviously be provided and a comprehensive changeroom service supplying each operator daily with a personal complete set of clean working clothes.

In certain parts of the plant the use of respirators is obligatory. The best type of respirator must be used with an adequate cleaning service. In view of the difficulty in getting operatives to wear respirators, particularly under hot conditions, reliance on their use should be minimised.

One important aspect is a policy of plant cleanliness. Every care should be taken when handling toxic-containing materials to avoid escape of dust, and any spillage on roads or elsewhere must be immediately moistened and removed.

Biological monitoring of operatives

In the UK it is a statutory requirement that every operative exposed to lead should be examined every three months by a specifically appointed medical officer, and determination made of blood lead content, and (not obligatory) of urinary coproporphyrin. If blood lead values exceed 80 mg/100 ml of whole blood, the operative must be withdrawn from exposure and not returned until the blood lead level content has dropped to below average values. The frequency of examination should be increased if blood lead levels vary appreciably or are consistently in the range 70–80 mg/100 ml. At Avonmouth in 1979 the blood lead level of operatives on the blast furnace plant varied between 45–50 mg/ 100 ml [4]. This would seem to be a reasonably good performance.

The Future

As was stated above, there is some confusion in industrial countries between the philosophies employed and the standards proposed for pollution control. In the EEC and other countries the complete legislation has not yet been finalised. The situation is one of extreme importance to the future of the zinc industry.

In the UK, pollution is controlled by the Alkali Inspectorate, and a responsible attitude has been taken, based on the experience of the 100 years it has been in existence. In some other countries the situation is different and standards are proposed which it is doubtful that the zinc industry could meet. There is also some doubt whether adequate medical evidence is available to justify them.

The question is a difficult one. Pollution cannot be condoned, but it is not possible to run an industrial society without causing some harm to the environment. There is no doubt that if real damage is still being caused, after the plants involved have done all that is possible to do to prevent it, then operations should not continue; but before such a serious step is taken the medical evidence should be irrefutable. Industry should be well aware of its responsibility and must do everything feasible to reduce environmental damage to a minimum, but the assessment of such damage should be firmly based on scientific data.

The question is highly emotive and it must be approached rationally. In the meantime industry should be given some protection from the 'strident polemics of pressure groups' [1] and the overzealousness of bureaucrats.

REFERENCES

[1] Barbour, A. K., European Economic Community regulations as applied to
     the lead and zinc industries, *AIME Lead-Zinc-Tin-'80 Symposium,* p. 592.

[2] Pearce, F., *New Scientist,* Vol. 103, No. 1420, p. 2.

[3] Atkins, M. H. and Lowe, J. F., *The Economics of Pollution Control in the
     Non-Ferrous Metals Industry,* p. 173, Pergamon Press, 1979.

[4] Robson, A. W., *Proc. Third Amax Energy Conservation Convention,*
     Indianapolis, Vol. 1, Paper C6, p. 23, Sept. 1980.

10

The physical metallurgy of zinc

10.1 THE CRYSTALLOGRAPHY OF ZINC

Zinc crystallises in the hexagonal system (see Fig. 36), the most generally accepted values of the lattice constants a and c being 0.2664 nm and 0.4947 nm respectively. Whilst the crystal structure is considered to be of the close-packed hexagonal type, the axial ratio c is 1.856, considerably greater than the theoretical value, 1.633, calculated for this system. As required by the close-packed hexagonal system each zinc atom has twelve near neighbours, but six are at one distance, 0.2664 nm, and six are farther away at 0.2907 nm. Thus the bonds in the hexagonal basal layers are appreciably stronger than those between layers, and this explains much of the behaviour of the metal under deformation, and the anisotropy of the zinc crystal.

Single crystals of zinc can be grown by a number of techniques, and the way such crystals deform under stress has been studied in considerable detail. Since atoms in the basal plane are closer to each other than to those in adjoining layers, bonding between basal planes is relatively weak, and when accommodating itself to stress, the lattice tends first to slip along such a plane. During such slip there is movement of part of the lattice along a basal plane so that the basic crystal structure is maintained. At higher than room temperatures slip may also occur along a prismatic (1010) plane.

Another major form of accommodation is twin formation which can occur readily along one of the (1012) pyramidal planes. In twin formation the basal plane of the twin forms at an angle of 94° to the original, and deformation due to twinning gives a reduction of 6.75 per cent in thickness measured perpendicular to the original basal plane. Once twinning has occurred, the orientation of the new basal plane is such that slip can take place more readily, and further deformation can be accommodated. Whilst the behaviour of polycrystalline material is complicated by the restraints imposed by the grain boundaries, slip and twinning are the two main methods of deformation, although two other associated types of lattice distortion, known as kinking and accommodation bending, can occur.

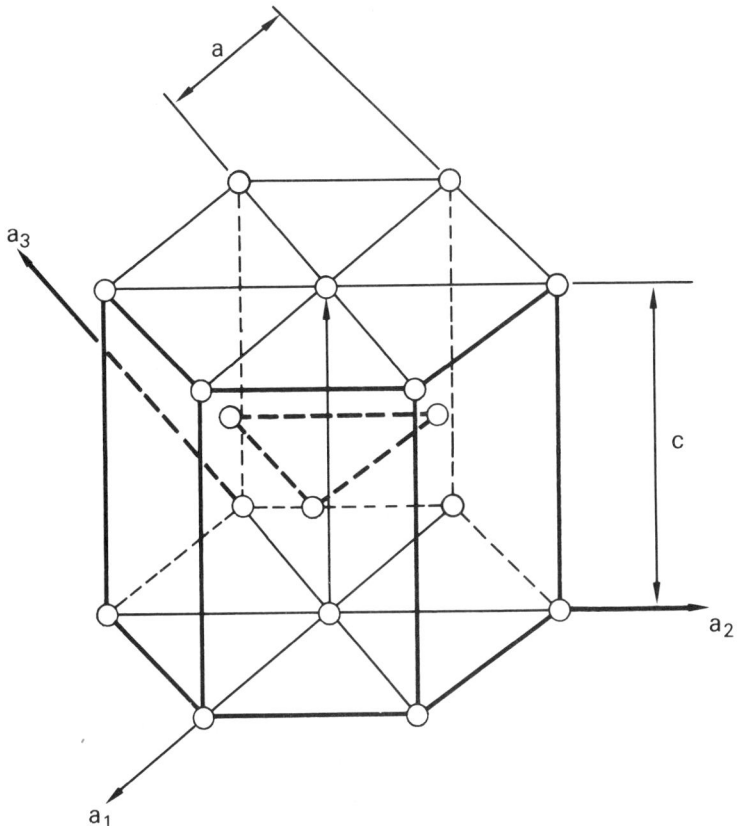

Fig. 36 – Hexagonal close-packed structure of zinc crystals.

A factor of great importance in the fabrication of zinc and the properties which can be obtained is the fact that the atomic mobility of atoms within the lattice is high at normal processing temperatures and recrystallisation takes place readily. Pure zinc recrystallises rapidly after deformation at room temperature, but with certain alloy additions forming solid solutions, such as cadmium or copper, somewhat higher temperatures are required. Thus, whilst with most other metals cold working produces strain-hardening with improved physical properties, with unalloyed zinc, recrystallisation gives rapid and almost complete relief from the strains imposed during fabrication, and little hardening occurs.

Whilst the ability of zinc to recrystallise readily leads to elimination of the structural hardening effects during deformation, the extent of this recovery is dependent on time and, thus, on the rate at which the stress is applied. Owing to this recovery mechanism and the accompanying reduction in the amount of

strain-hardening, the creep-resistance, or the ability of zinc to withstand deformation under prolonged light loads, is low. This has been one of the most serious
drawbacks to the use of zinc as an engineering material, although the development in recent years of zinc alloys containing small additions of titanium and
copper with resistance to creep many times that of unalloyed zinc goes some
way towards rectifying this disability.

In estimating the strength of a metal, the critical shear stress, or that stress
at which elastic deformation ceases and irreversible lattice movements occur,
is an important parameter. With zinc, the critical shear stress is low and is
dependent on the rate of application of the stress and also on the temperature.
Single crystals of zinc have been shown to bend in time under their own weight,
and the critical resolved shear stress under these conditions must be as low as
8 g/mm^2. (0.0785 MNm^{-2}.)

10.2 ZINC AS AN ALLOYING ELEMENT

The main characteristic of zinc as an alloying element is that its solid solubility
in other metals is very limited and, additionally, few metals have any appreciable
solid solubility in zinc. It does not form a continuous series of solid solutions
with any other metal, although in aluminium the solid solubility can reach 80 wt
per cent, in copper 39 per cent and in iron 20 per cent. It is also soluble to a
lesser degree in magnesium, manganese, nickel, cobalt and gold, but in most
other metals the solubility is low at less than 3 per cent.

Gold and silver dissolve in solid zinc to the extent of about 10 per cent,
which is the basis of the Parkes process used to remove these metals from molten
lead. Cadmium and copper have a maximum solubility of 3 per cent, but that of
other metals, other than manganese and aluminium (less than 1 per cent), is very
low.

The low creep-resistance of unalloyed zinc sheet is one of its most serious
defects and restricts its application. During recent years a considerable amount
of work has been carried out on this problem, and much is now known about
the adjustments of the zinc lattice under stress and the mechanism of the creep
reaction, which occurs mainly through grain boundary migration. For some time
it had been known that small additions of a number of elements which had some
solid solubility in zinc, such as copper, chromium and cadmium, gave a limited
improvement, but a major advance was made in 1944 by the New Jersey Zinc
Company when it was shown that with suitable reduction and annealing techniques, additions of titanium could give a marked increase in creep-resistance.
As a result of this work a number of titanium-containing alloys have been
advocated, but the optimum properties appear to be developed with additions
of 0.05–0.2 per cent titanium and 0.5–1 per cent copper [1].

The alloy systems of principal interest are:

(1) Zinc–aluminium, which at 4 per cent aluminium forms the basis of the zinc die-casting alloys.
(2) Zinc–copper, where with zinc up to 45 per cent brasses are formed.
(3) Zinc–iron, this system being the basis of the corrosion protection of steel by galvanising.
(4) Zinc–lead, which plays an important role in pyrometallurgical extraction processes.

In addition, the phase system of zinc with magnesium, titanium and cadmium have significance in commercially available alloys.

Zinc–aluminium

The constituional diagram shown in Fig. 37 indicates conditions favourable for the production of useful alloys at both ends of the system, i.e. extensive solid solubility of zinc in aluminium and limited solid solubility of aluminium in zinc (1.14 wt per cent). The use of aluminium as an essential constituent of zinc-base die-casting alloys at compositions near the eutectic value of 5 wt per cent is discussed in Chapter 11.

Zinc at up to 6 wt per cent is also an essential component of certain high-strength wrought-aluminium alloys, and at lower concentrations in some of the cast aluminium alloys. Alloys of zinc with aluminium can show the effect known as superplasticity and attempts are being made to commercialize an alloy containing 20 wt per cent aluminium.

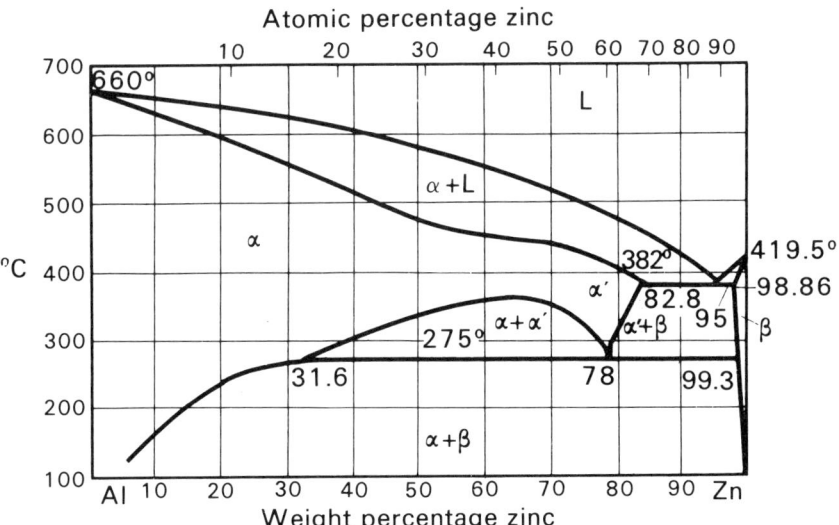

Fig. 37 – Zinc–aluminium equilibrium diagram.

Zinc–aluminium alloys of commercial use in gravity casting have been developed recently, e.g. ZA 8, ZA 12 and ZA 27. Their properties are discussed in Chapter 11.

The addition of approximately 0.1 per cent of aluminium to zinc used for galvanising greatly reduces the formation of the brittle intermediate zinc–iron compound layer, and the hot-dipped coatings formed in baths with such aluminium additions are much more ductile than those without aluminium. Such additions are always made in the continuous strip galvanising process. The presence of similar quantities of aluminium produces a more even rate of corrosion of zinc in anodes used for protection and thus improves their effective life. Aluminium if present in excess of 0.005 per cent can stimulate intercrystalline corrosion in rolled zinc — an effect which is increased if cadmium, tin or copper are also present.

Zinc–copper

As can be seen from the phase diagram, Fig. 38, zinc alloys readily with copper in all proportions, but alloys containing up to about 45 per cent zinc are in

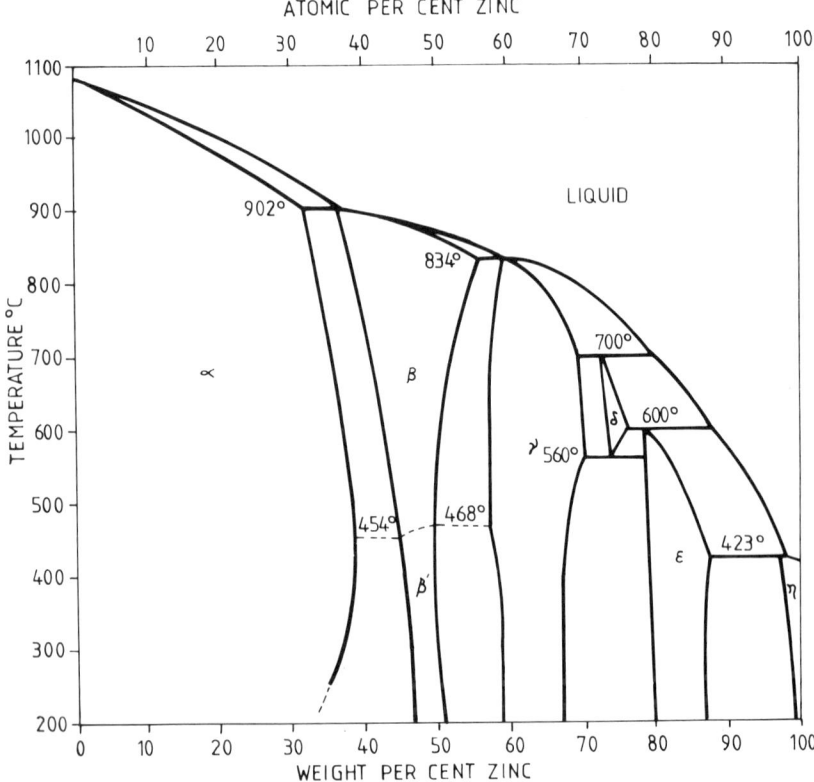

Fig. 38 – Zinc–copper equilibrium diagram.

commercial use as 'brass' and range in colour from red through yellow to gold as the amount of zinc increases. Two main phases are involved, the α-phase which is face-centred cubic with a maximum zinc content of 39 wt per cent at 454°C, and the β-phase which is body-centred cubic and occurs at higher zinc concentrations. As the amount of zinc in the alloy increases, the solidified and cooled alloys change from α- to mixed α + β-phases at around 40 per cent zinc, and to the β-phase alone approaching 50 per cent zinc. Alloys composed entirely of the α-phase are characterised by good adaptability to cold processing. The β-phase is plastic at red heat but on cooling becomes brittle. In combination with the α-phase around 40 per cent zinc, hot plasticity is followed on cooling by reasonable cold working properties.

The intermediate phases beyond β including γ are compound-like associations of zinc and copper atoms and are hard and brittle.

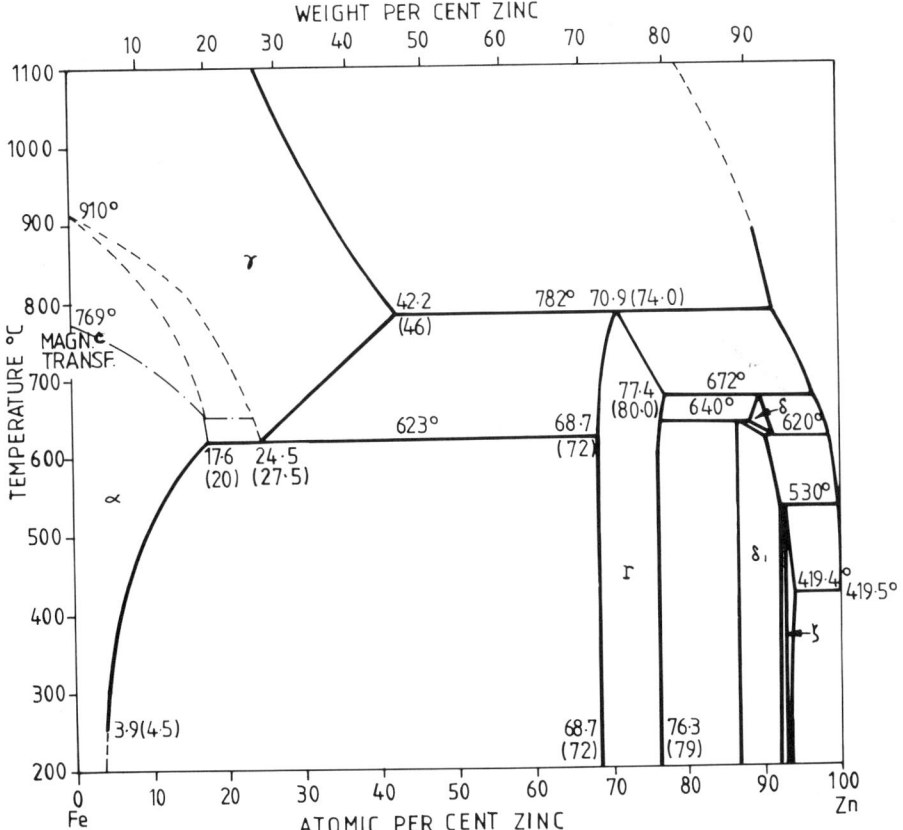

Fig. 39 – Zinc–iron equilibrium diagram.

Full details of the brasses are to be found in the companion book on copper in this series [1].

At the zinc-rich end of the diagram, the maximum solid solubility of copper is 2.7 per cent. The presence of copper raises the recrystallisation temperature of zinc, and therefore increases tensile strength, hardness and creep-resistance. Improved wrought zinc alloys containing 0.5–1.0 per cent copper and 0.1 per cent titanium are discussed in Chapter 11, Section 11.3, page 172.

Zinc—iron

The full phase diagram is shown here Fig. 39, but the zinc-rich end of the diagram is shown in Chapter 12, page 202, along with the discussion on the corrosion-protection ability of zinc coatings on steel. When iron is present above 0.001 per cent, its presence can be detected micrographically by the appearance of an intermediate compound, possibly $FeZn_7$. The presence of this compound is frequently beneficial, since it can be used to control the grain size and therefore the strength of rolled zinc, but, if its concentration exceeds 0.008 per cent, brittleness and loss of ductility develop. Iron has a considerable effect on the chemical properties of zinc, and if present at concentrations above 0.0014 per cent, it causes the formation of films of corrosion products of high electrical resistivity on the zinc anodes used to protect steel structures in sea water, although this effect can be minimised by the presence of 0.1 per cent of aluminium. If

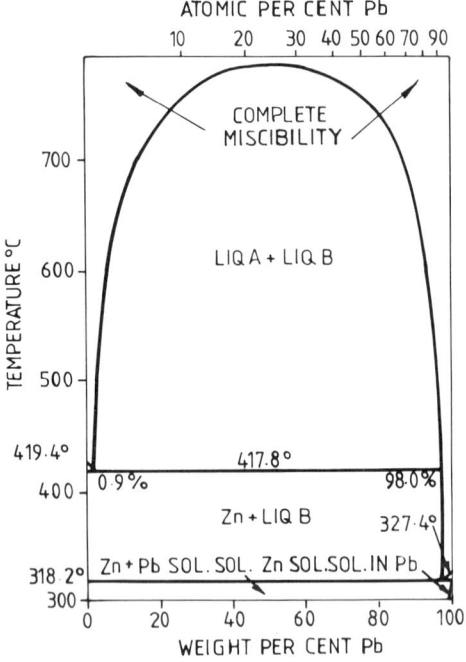

Fig. 40 – Zinc–lead equilibrium diagram.

present above 0.0025 per cent, iron increases the rate of attack of acids on zinc. For use in primary electric cells the iron content should not exceed 0.004 per cent, otherwise the shelf-life of such cells deteriorates.

Zinc—lead

Lead is virtually insoluble in solid zinc, and even with high purity metal containing less than 0.001 per cent lead, it can be detected as beads at the grain boundaries. As a consequence it has little effect on the mechanical properties of zinc, except that owing to its low melting point and its concentration at grain boundaries, its presence reduces the temperature at which zinc can be worked. For certain applications, such as dry cells, lead reduces the rate of corrosion of zinc and is added to improve shelf-life. In galvanising, lead accentuates the formation of spangle on the finished product, and is frequently added for this purpose.

The zinc—lead system is reproduced here (Fig. 40) to underline the significance of the method of continuous separation of liquid zinc from solution in liquid lead in the condenser system of the blast furnace process. (See Chapter 5 page 74.)

Zinc—Magnesium

The phase diagram is shown in Fig. 41.

Zinc alloys freely with magnesium, with a maximum 8.4 per cent solid

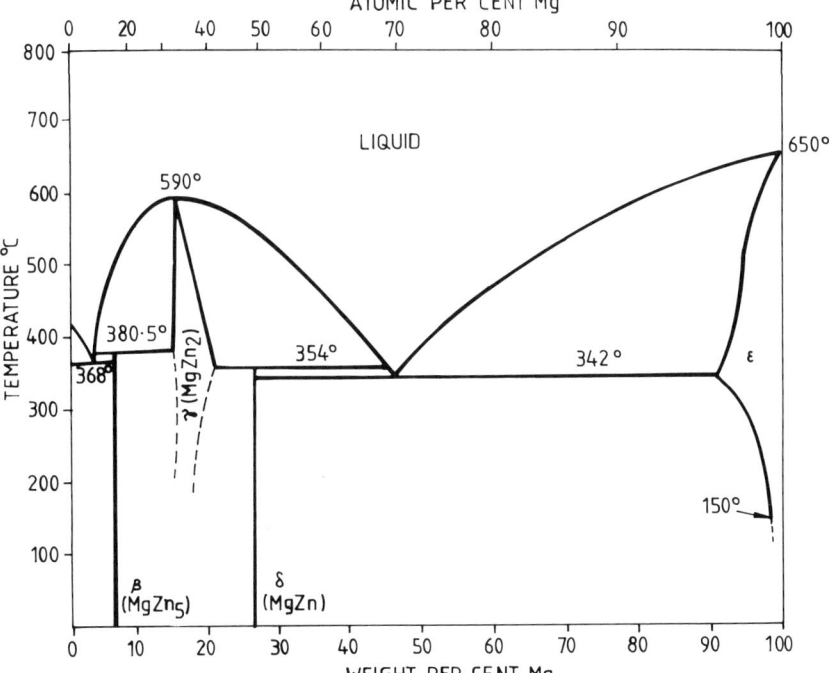

Fig. 41 – Zinc—magnesium equilibrium diagram.

solubility of zinc in magnesium but no appreciable solid solubility of magnesium in zinc.

Zinc is used at low percentages in a number of wrought and cast magnesium alloys, generally in combination with a larger amount of aluminium.

Zinc–titanium

The zinc-rich end of the phase diagram is shown in Fig. 42.

Titanium has a limited solubility in zinc, forming at about 0.19 per cent titanium a eutectic of zinc and the intermetallic compound $TiZn_{15}$. This compound has a markedly beneficial effect in decreasing the grain size of cast zinc and restricting grain growth at elevated temperatures. Whilst having little effect on the tensile strength and hardness of rolled zinc, it improves creep-resistance considerably, and is added for this purpose.

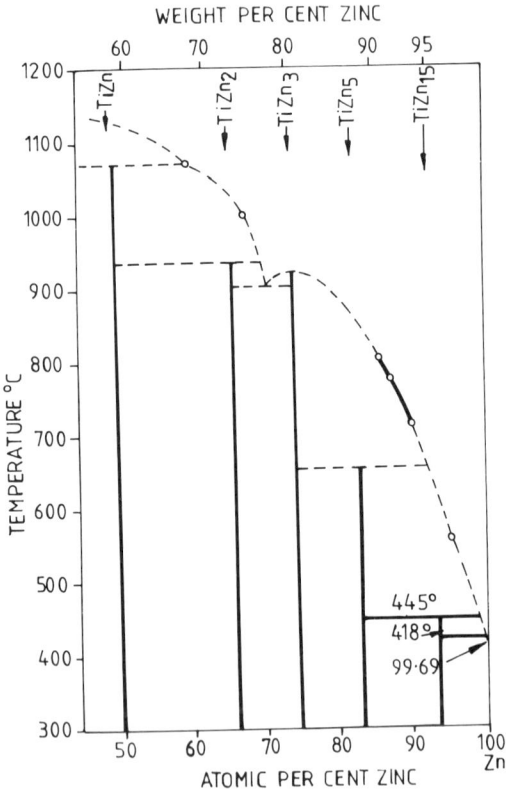

Fig. 42 – Zinc–titanium equilibrium diagram.

Zinc–cadmium

Zinc and cadmium form a simple eutectic system with some solid solubility of each component in the other, as can be seen in Fig. 43. In fact, cadmium is one of the few metals which is held in solid solution in zinc. Up to 0.01 per cent it has little effect on the microstructure or properties of cast zinc, and is sometimes added to zinc which is to be rolled since it raises the recrystallisation temperature and improves strength, hardness and creep-resistance.

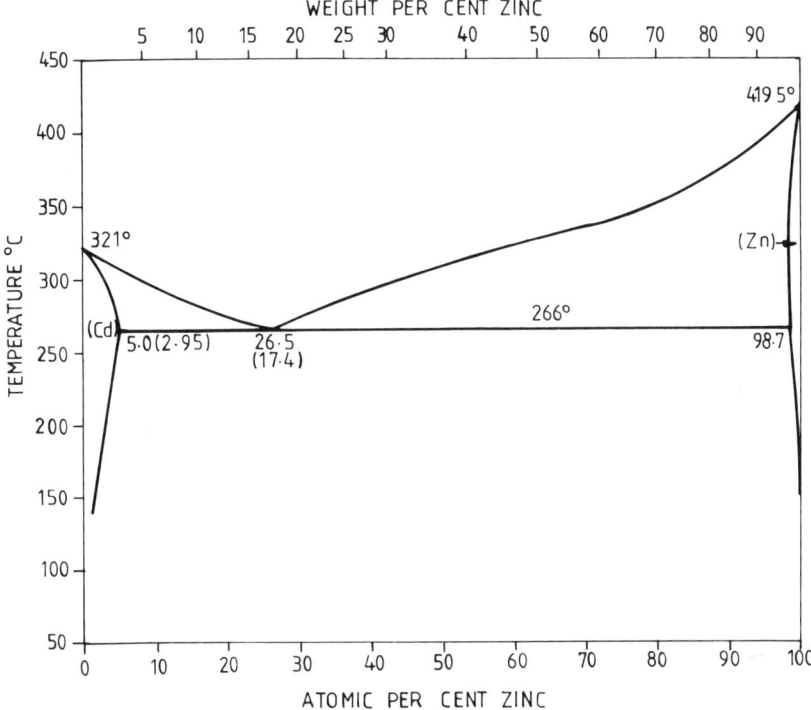

Fig. 43 – Zinc–cadmium equilibrium diagram.

Zinc–tin

Zinc and tin also form a simple eutectic system similar to that of zinc–cadmium, but with very limited solid solubility. It is now considered that the solid solubility of tin in zinc is less than 0.001 per cent, and above this a eutectic phase at 9 wt per cent zinc, melting at 200°C, appears at grain boundaries, which is troublesome during rolling. The presence of tin is very harmful in the zinc–aluminium alloys and must not exceed 0.001 per cent because it promotes intercrystalline corrosion in humid conditions.

At the tin-rich end of the system, the solid solubility of zinc in solid tin is probably about 1.0 wt per cent.

REFERENCES AND FURTHER READING

[1] West, E. G., *Copper and its Alloys,* Ellis Horwood, Chichester, pp. 98–105, 1982.

Additional information may be obtained from several reference books including:
Smithells, C. J. (editor), *Metals Reference Book,* Butterworth, London, 1967.
Hansen, M., and Anderko, K., *The Constitution of Binary Alloys,* McGraw-Hill.

11

Applications of zinc and zinc alloys

As would be expected from the relatively low melting point, zinc, and the alloys of which it forms a major constituent, cannot be used for highly stressed applications. Nevertheless the zinc–aluminium alloys, with their good castability, fill a gap between plastics and the stronger but higher melting point aluminium alloys. Consequently one of the main outlets for zinc is in the form of die castings used in the automobile and consumer durables industries.

Another major use for zinc is the series of alloys with copper to form the brasses and nickel silvers. By varying the zinc content from 15 per cent to 40 per cent, alloys with a wide range of strength and ductility can be produced.

An important outlet for the metal itself is in rolled form, zinc sheet being used to a considerable extent for roofing and cladding, especially in certain parts of France and Germany. It is handicapped for this purpose by a low resistance to creep, although this can be overcome to a certain extent by small additions of copper and titanium.

The most important use for zinc, however, is for the protection of steelwork against corrosion, where its relatively high corrosion-resistance and electro-chemical properties give it an almost unique position in this field. A number of methods of coating iron with zinc are used, the main one being hot-dip galvanising, but electrogalvanising and zinc spraying are also applied to a considerable extent.

11.1 ZINC CONSUMPTION BY INDUSTRIES

The proportion in which zinc is used in these main fields varies from country to country. The pattern of distribution in 1982 for the United Kingdom, United States, Germany and Japan is shown in Table 17.

The figures in Table 17 refer to the total consumption of metal in the consuming industries listed and they include virgin zinc and a proportion of re-circulated metal recovered from scrap and secondary materials. The proportion of recovered metal varies, being high in zinc dust and oxide, as well as in brass production, but low in die castings. Overall the proportion of circulating zinc is approximately 15 per cent of the virgin metal consumption, which is much lower than for aluminium (48 per cent), copper (61 per cent) and lead (42 per cent). In the interest of conservation this proportion should and could be increased, but it is possible to do this only to a limited extent, since much of the zinc used in steel protection, for zinc oxide and for zinc-dust production cannot be reclaimed.

The applications of zinc and the zinc-base alloys are dependent, of course, on their physical and mechanical properties. A brief account is given in Chapter 10 of the crystallography of zinc and its properties as they affect its industrial usage.

Table 7

Consumption of zinc by industries 1982 ('000 tonnes)

	UK	%	US	%	Germany	%	Japan	%
Brass	59.3	24.8	73.2	10.5	81.8	21.6	101.0	13.9
Galvanising	83.5	34.8	310.2	44.5	137.0	36.2	412.9	56.7
Rolled zinc	14.1	5.9	37.1	5.3	66.1	17.5	29.3	4.0
Die casting	37.5	15.7	157.8	22.6	71.0[a]	18.8	108.7	14.9
Zinc oxide	18.7	7.8	17.6	2.5	9.7	2.6	Chemicals	
							38.9	5.3
Zinc dust	9.8	4.1	–	–	–	–		
Miscellaneous	16.4	6.9	101.4	14.5	12.9	3.4	37.0	5.1
Total	239.3		697.3		378.5[b]		727.8	

[a] Includes zinc oxides.
[b] Slab zinc + remelted only.

11.2 ZINC SHEET AND STRIP

Zinc can be fabricated by almost all the processes used in metal forming, and considerable quantities are rolled, extruded and drawn, although the peculiarities of the close-packed hexagonal lattice of zinc require some modification of the techniques commonly employed. As unalloyed zinc recrystallises at room temperature the effects of work hardening are quickly removed and it is not possible to harden zinc appeciably by working.

Rolling is the most commonly applied method of fabrication, and rolled zinc satisfies many uses, in spite of its rather low mechanical properties. Among the applications of wide strip, roofing is the most important, as illustrated in

Fig. 44. Whilst previously zinc of all grades was produced in rolled form, the tendency today is to use specified compositions with small alloy additions aimed to improve the grain size and properties, in particular, creep-resistance. A number of proprietary compositions are on the market with additions of copper, manganese, magnesium, aluminium, chromium and titanium, but the most widely used alloy contains copper and titanium, developed originally by the New Jersey Zinc Company.

Fig. 44 – Zinc rolled cap roofing on the Magistrates' Court Building, Oxford.
(Courtesy of Zinc Development Association.)

Although the amount of zinc sheet produced in continuous mills is increasing rapidly, a proportion is still produced by the old established pack rolling method. In this process zinc is first cast into slabs from 20–100 mm thick in open or closed book-type moulds made of cast iron. Open moulds are generally water-cooled and use heated tops to reduce solidification shrinkage cavities. Book moulds are air-cooled and tend to produce a better quality ingot. The continuous casting of slabs is now widely practised since the greater uniformity of grain size and uniformity of crystal structure is of considerable advantage.

The ingots are first reduced in a slab mill. A cast slab of unalloyed zinc is composed mainly of coarse columnar grains orientated so that the base plane is parallel to the column axis, and hence normal to the slab surface. The first stage in the breakdown of this structure by rolling is the compression of these columns and this is carried out by basal plane bending since slip cannot easily occur. Reduction per pass through the mills is usually limited to below 10 per cent at these early stages and roll pressures are high. The operation is carried out at temperatures between 160°C and 250°C so that the maximum amount of slip is encouraged and cracking by cleavage is avoided. Once the original columnar structure is broken down, deformation occurs by both slip and twinning, and reductions of 30 per cent or more can be applied at each stage.

After the breaking-down operation in the slab mill, the sheets, which now are approximately 3 mm thick, are cut into lengths equal to the width of the finishing rolls, and placed together in packs, the number varying from 2 to 40 depending on the gauge required in the finished product. The stacked sheets are rolled together in the finishing mill in a direction now at right angles to that used for slab rolling. By cross rolling, the sheets are worked in two directions, reducing to some degree the effects of anisotropy. With care, good quality sheets can be produced by the pack rolling method, but considerable variation in properties of the sheet produced can occur.

The continuous casting and rolling of zinc

Pack rolling for zinc sheet production involves extensive manual handling. In recent years much attention has been given to the development of methods for casting and rolling zinc continuously. Mills producing strip and even foil up to 750 mm wide have been in use for many years but only recently has continuous rolling of wide strip been achieved, by the development of the Hazelett casting machine, shown diagramatically in Fig. 45.

The machine consists essentially of two continuous belts 1000 mm wide X 7 m long supported and driven through a system of pulleys. These form the walls of the mould, which is closed at the sides by two sets of small steel blocks linked together by cables, forming two endless chains sandwiched between the belts. Molten metal is fed from a tundish or funnel into the cavity between the belts, which are cooled by water sprays. The belts are rotated at a controlled speed, and as the metal solidifies the slab is drawn downwards, thus producing a

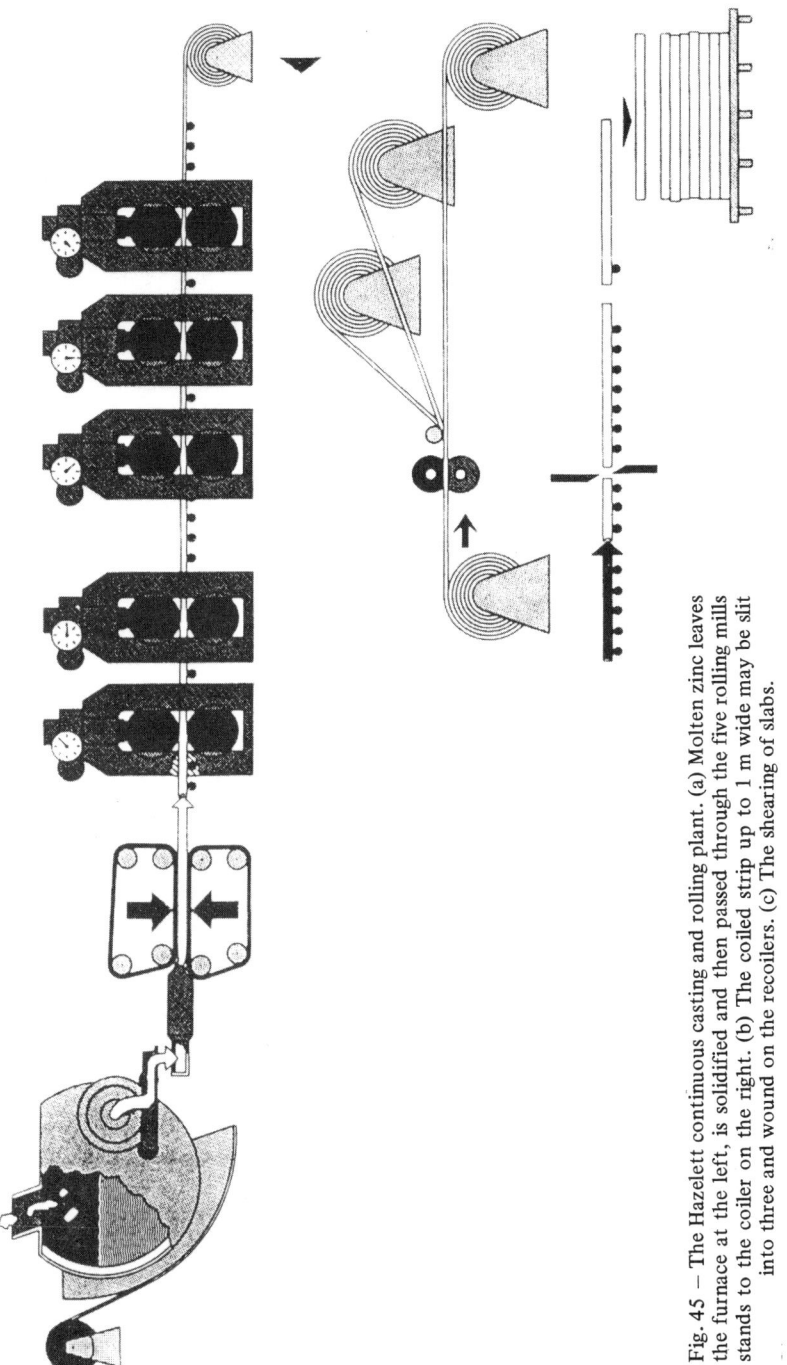

Fig. 45 – The Hazelett continuous casting and rolling plant. (a) Molten zinc leaves the furnace at the left, is solidified and then passed through the five rolling mills stands to the coiler on the right. (b) The coiled strip up to 1 m wide may be slit into three and wound on the recoilers. (c) The shearing of slabs.

continuously cast ingot, the thickness of which can be varied from 10 to 75 mm by adjusting the belt separation. The width of the slab or strip is 1000 mm. Owing to the continuous nature of the casting operation, and particularly when alloyed zinc is treated, the grain size of the solid slab is much finer than when static moulds are used, and subsequent reduction is therefore much easier, giving a more uniform product.

After leaving the casting machine the slab is cooled by water sprays to 180–240°C and then fed into the mill, which is generally of the four high roll type, where a reduction of 60 per cent is taken in one pass. The strip is then coiled and, when the casting run has been completed, is fed back through the mill and rolled at a temperature of 80–90°C, with reduction up to 50 per cent, to give the required final thickness, finish and properties.

There are 1-m-wide Hazelett mills for rolling zinc in France and Germany, and narrower mills using the same principle are operated by the Ever Ready Company in the UK, in Germany and the USA.

Forming of zinc sheet

Zinc of high purity is ductile, but commercial grades tend to be brittle and are difficult to work in the cold. Malleability increases rapidly when the temperature is increased, and if zinc sheet is to be flanged, or worked in other ways, its temperature should be raised to 40–50°C.

Battery cans

One of the major outlets for zinc sheet is in the production of dry batteries of the Leclanché type, widely used in portable electrical equipment. The zinc forming the anode electrodes of these cells generally consists of a cylindrical case, which also acts as the container for the cell. Such cans are usually made by impact extrusion from blanks or calots stamped out of rolled strip. Larger cells are built up from layers of zinc with carbon cathodes deposited on one side.

Figure 46 shows typical impact-extruded battery cans with the calots from which they were formed. The zinc used contains

0.8%–1.0%	lead
0.05%	cadmium

and the main impurities are generally held below

0.002% iron
0.005% arsenic
0.006% calcium

An upper limit of 0.004 per cent is fixed for the iron content of the zinc used for battery cans, to reduce corrosion, and increase shelf-life. Copper is held below 0.001 per cent. Cadmium up to 0.05 per cent is sometimes added to facilitate impact extrusion.

Fig. 46 — Typical impact-extruded zinc battery cans and the calots from which they are made. The round calot is about 5.5 mm thick and the narrow can has sides about 0.4 mm thick and is 50 mm high. The hexagonal calot is about 3 mm thick and the large can has sides about 0.4 mm thick and is 57 mm high.

Zinc coins

In 1982 the US Government adopted 1 cent coins stamped from a rolled zinc 0.8 per cent copper alloy, plated with about 6 μm of copper. The coins are virtually indistinguishable from the former copper coins and considerably cheaper. Over 30,000 tonnes of zinc per year is needed for this application.

Lithography and photoengraving

Some zinc is still used in photoengraving and lithography. Sheet for such purpose is generally pack rolled to give a high surface finish and small grain size. High purity zinc, low in lead, is used, but alloying additions (generally copper or magnesium) are sometimes made to give a fine grain and to improve stiffness. Although in the past zinc has been widely used in these applications, it is now being replaced by plastics and magnesium.

Zinc wire

Zinc wire can be made by casting cylindrical billets, which are subsequently extruded, rolled and finally drawn. Nowadays it is more common to use the Properzi continuous casting, rolling and drawing process. Molten zinc is poured into the triangular space formed between a steel belt and a groove in the rim of a cast iron wheel about 2 m in diameter. The triangular-sectioned cast rod is pulled away from the casting machine into a series of pairs of rolls which develop a circular section ready for wire drawing. Zinc of 99.99 per cent purity is used to make wire for zinc spraying, and an alloy containing a small addition of manganese — about 0.2 per cent — is used to make the wire which is incorporated as a mesh in vehicle brake and clutch linings, where it improves thermal conductivity and eliminates squealing. The manganese gives the zinc enough strength to avoid breakage during fabrication into mesh.

11.3 ZINC–TITANIUM–COPPER ALLOYS

The development of the zinc–titanium–copper alloys and techniques of continuous casting and rolling have produced what can be considered as virtually a new material. Commercially quoted figures for the mechanical properties of continuously rolled sheet of zinc plus 0.1 per cent titanium and 0.5 per cent copper compared with aluminium, brass and copper are given in Table 18.

Table 18

Comparison of mechanical properties of
copper–titanium–zinc and other materials

	Zn–Cu–Ti alloy	Aluminium 99.5%	70/30 Brass annealed	Copper cold-rolled
Ultimate tensile strength (MNm^{-2})				
parallel to rolling	162	140	309	371
normal to rolling	193			
% elongation				
parallel to rolling	40	10	70	4
normal to rolling	20			
Hardness (V.P.N.)	58	42	66	115
Young's modulus (GNm^{-2})	96.0	69.6	103.4	124.1
Density (gm/cm^3)	7.2	2.7	8.52	8.9

Figure 47, from Brown and Laurie [2], compares the relationship between inverse creep rate and applied stress for zinc and zinc alloys and compares them with aluminium. The considerable improvement brought about by the addition of copper and titanium to zinc is clearly shown. Work carried out in Germany by Pelzel [3] and his colleagues on the influence of copper and titanium (as well as a number of other elements) on zinc showed that with suitable extrusion and annealing techniques, creep rates as low as 1–2 per cent per year under stresses up to 1200 kg cm^{-2} (117.7 MNm^{-2}) could be obtained. With unalloyed zinc the same deformation would be reached in a few days. With rolled sheet the improvement is of the same order.

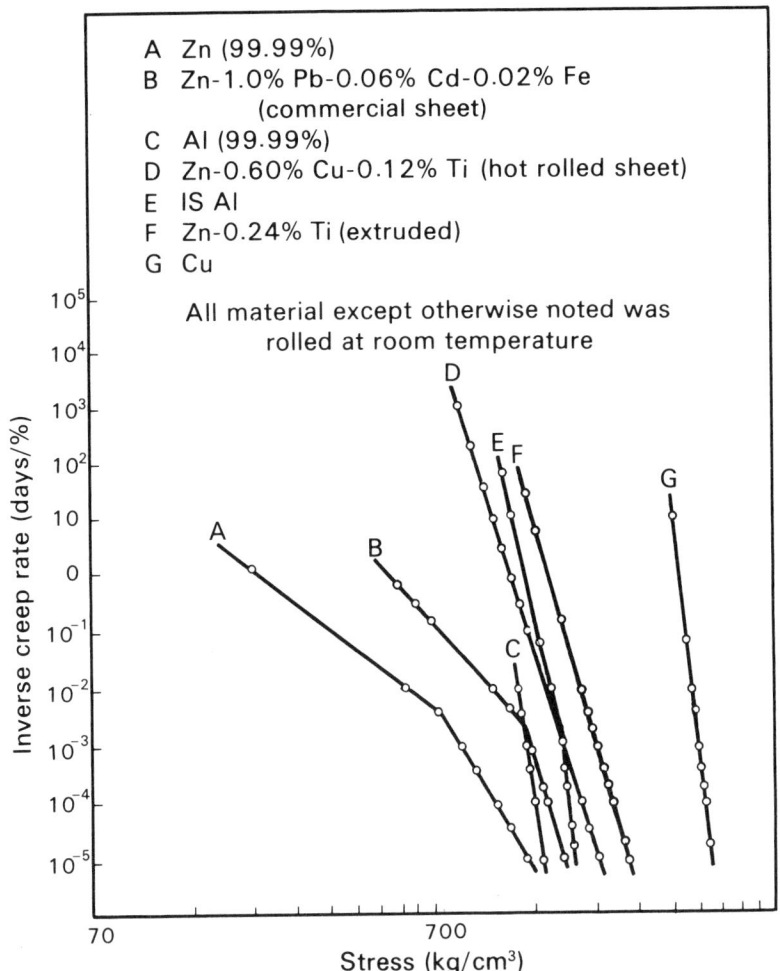

Fig. 47 – Inverse creep rate as a function of applied stress for zinc and zinc alloys, as compared with aluminium and copper. A Zinc 99.99% purity. B Commercial sheet zinc–1.0% lead–0.06% cadmium–0.02% Iron. C Aluminium 99.99% purity. D Hot-rolled sheet zinc–0.60% copper–0.12% titanium. E Commercial purity aluminium sheet. F Extruded zinc–0.24% titanium alloy. G Copper. (All material except otherwise noted was rolled at room temperature.)

In one of the earliest published studies of the zinc–copper–titanium rolling alloys, Rennhack [4] showed that the microstructure of cast alloys consisted of sub-cells of a zinc-rich solid solution surrounded by a eutectic of this solid solution and plate-shaped particles of the inter-metallic compound $TiZn_{15}$. During the early stages of rolling, these compound particles gradually orientated to form 'stringers' which ultimately formed fibres. Reorientation of the fibres was complete after 50 per cent reduction, when they became aligned parallel to the direction of rolling.

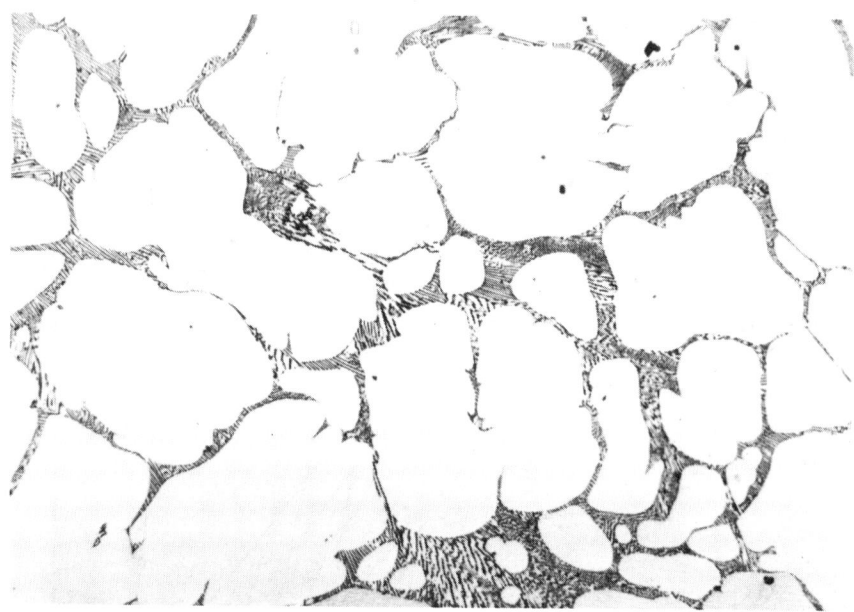

Fig. 48 — Zinc–copper 1% — titanium 0.1%: as cast. (Etched.)
(Magnification ×650.)

Fig. 49 — Zinc–copper 1.0%–titanium 0.1%: rolled and annealed at 400°C.
(Etched.) (Magnification ×650.)

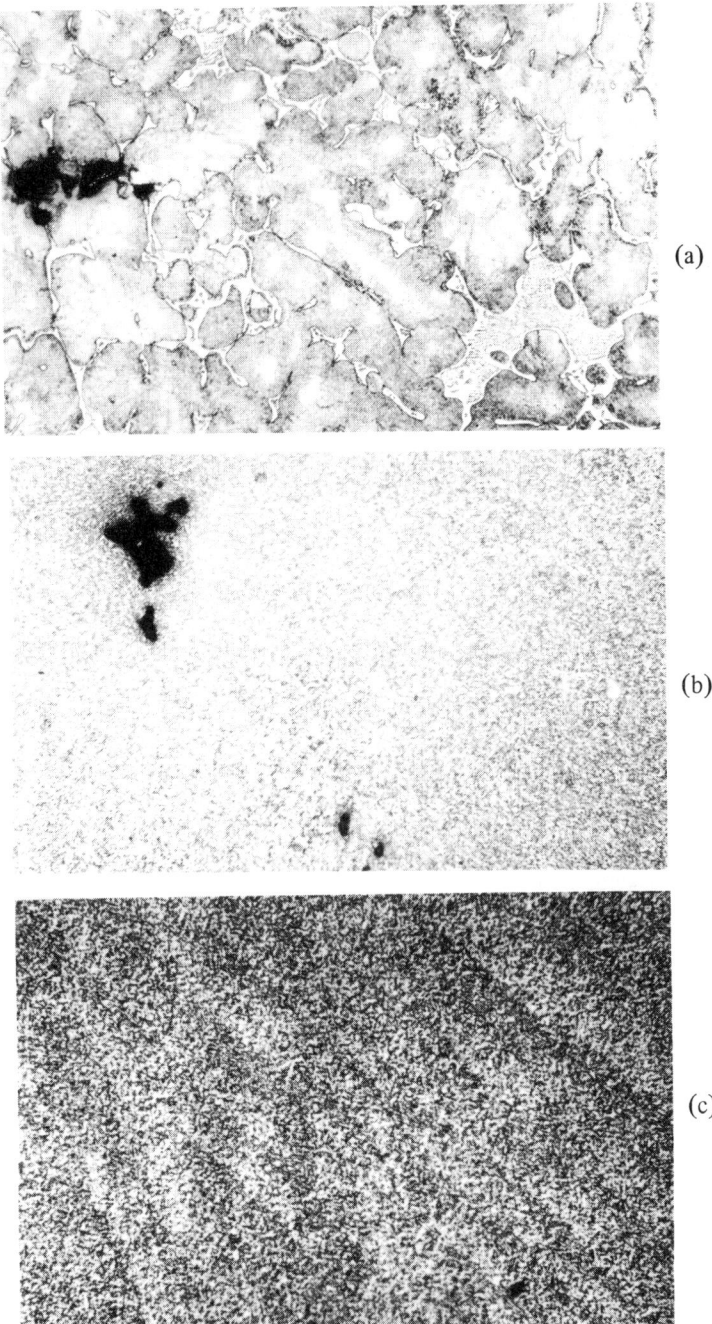

Fig. 50 – Superlastic zinc alloy. (a) As cast, 80% zinc, 20% aluminium. (b) Cast and quenched from 350°C, 80% zinc, 20% aluminium. (c) Cast quenched and rolled at 250°C, 80% zinc, 20% aluminium, (All: magnification ×250.)

Superplastic zinc

A recent commercial development is the application of 'superplasticity' which is exhibited by a number of zinc and other alloys. Under the appropriate conditions, the material becomes exceptionally soft and ductile, and under low stresses, extensions of the order of 1000 per cent can be obtained without fracture. A common requirement is that the grain size should be extremely small – of the order of 1 μm, less than one tenth that of metals in the conventional state [6, 7].

Extremely fine-grained two-phase structures can be developed in zinc–aluminium eutectoid alloys (approximately 22 per cent aluminium) by rapid quenching from above the eutectoid temperature of 275°C, or by rolling to appreciable degrees of deformation at somewhat lower temperatures, as can be seen from the micrographs (see Fig. 50). If sheet in which this fineness of grain size has been developed is heated to just below the transformation temperature (200–275°C), superplastic behaviour can be obtained. Because of the ductility and low strength in this condition, the material is comparable with thermoplastic polymers and can be moulded like them into complex shapes by vacuum- or blow-forming, stretch-forming, forging and other techniques; but when cooled to room temperature, like most metals, its strength and hardness are considerably greater than that of any thermoplastic.

Micrographic examination of superplastic zinc–aluminium sheet before and after severe superplastic deformation shows little distortion of the grains, and deformation appears to take place by slip of grains over each other. Thus, the very low strength and high plasticity in the superplastic state cannot be explained in terms of the normal plastic deformation of metals, but, for reasons which are uncertain at the moment, the strength across the grain boundaries is greatly reduced in the plastic range.

The development of these alloys offers the designer materials which can be moulded easily like thermoplastics at moderately elevated temperatures (200– 275°C) but which at room temperature have normal metallic properties, including strength, stiffness, conductivity and stability, which no plastic can equal. The tooling cost is generally considerably less than for conventional pressing, which is particularly beneficial for limited production runs. Commercial development has been hindered by difficulties in producing the strip economically in small quantities.

11.4 ZINC DIE-CASTING ALLOYS

For over 100 years, permanent iron moulds have been used for the repetitive production of numbers of metal castings, and with increasing mechanisation die casting has become an important production operation. Since the beginning of the century attempts were made to die cast zinc alloys, but in many cases rapid deterioration set in after a few months of service, and results were disastrous. The problem was studied by the New Jersey Zinc Company, and in 1923 it was discovered that the commercial alloys had failed owing to intercrystalline corrosion caused by impurities, such as lead or tin, which have almost negligible solubility in solid zinc. No deterioration occurred if the alloys were compounded of high purity zinc (99.99+ per cent) so that the offending impurities could be held to very low levels. The investigators soon put forward a series of alloys to which they gave the name Zamak (known as Mazak in the United Kingdom), which have not been improved upon, and which are still used today. In 1982, approximately 700,000 tonnes of zinc were die cast to the original Zamak composition.

In 1942 the British Standards Institution produced a specification, BS 1004, covering two of the Zamak compositions, which have the best all-round balance of properties, and which can be used for most applications. In the United States, the American Society for Testing Materials issued a specification, B-86, covering the same alloys, but which allowed more latitude in impurity content. Current versions of the specifications for the composition of the castings are given in Table 19. The very low tolerance allowed for lead, tin and cadmium means that every precaution must be taken to exclude these elements at all stages of alloying and casting, and particular care must be taken to ensure that worked scrap returned to the melting pots is uncontaminated.

Table 19

Zinc alloy die-casting compositions

		Alloy Zamak 3 (USA) Mazak 3 (UK)		Alloy Zamak 5 (USA) Mazak 5 (UK)	
		UK Specification BS 1004 A	USA Specification B-86	UK Specification BS 1004 B	USA Specification B-86
Aluminium	min.	3.8	3.5	3.8	3.5
	max.	4.3	4.3	4.3	4.3
Copper	min.	–	–	0.75	0.75
	max.	0.03	0.25	1.25	1.25
Magnesium	min.	0.03	0.02	0.03	0.03
	max.	0.06	0.05	0.06	0.08
Iron	max.	0.10	0.10	0.10	0.10
Nickel	max.	0.02	0.02	0.02	0.02
Manganese	max.	0.01	0.01	0.01	0.01
Lead	max.	0.005	0.005	0.005	0.005
Cadmium	max.	0.005	0.004	0.005	0.004
Tin	max.	0.002	0.003	0.002	0.003
Zinc		remainder	remainder	remainder	remainder

Effect of alloying elements

Aluminium

Aluminium is the main alloying constitutent and has a major effect in increasing strength and reducing grain size. If present in excess of 4.3 per cent, impact strength is reduced, and at 5.0 per cent the alloy becomes brittle; but should the aluminium content drop below 3.7 per cent, ease of casting and mechanical properties deteriorate. The composition therefore must be held within relatively narrow limits. With aluminium present, the attack on steel is greatly reduced, long die-life can be obtained and rapid-acting submerged-plunger-type ('goose-neck') machines are used.

Copper

Copper tends to increase tensile strength and hardness, but reduces impact strength and causes some dimensional instability on ageing. In consequence, the copper-free alloy A is used for most applications, and alloy B reserved for those applications where the highest tensile strength, hardness and castability are required.

Magnesium

Magnesium plays an important part in reducing the possibility of intercrystalline corrosion. If present above 0.06 per cent, impact strength and ductility are adversely affected and brittleness at elevated temperatures (hot shortness) may develop.

Iron

Iron has little effect on the properties of the die-cast alloy. Approximately 0.02 per cent is held in solid solution, but above this concentration iron tends to form a hard compound with the aluminium present, and can cause machining problems – e.g. drill breakage.

Lead and tin

Lead and tin above the limits given in the specification for either element are highly dangerous, since if they are exceeded, corrosion in humid atmospheres can spread rapidly along the grain boundaries, where the impurities are concentrated, and the strength of the casting is seriously reduced. The quantities which cause this are so small that, in the casting shop, extreme precautions are essential to avoid contamination, and only high quality scrap should be returned to the melting pots.

Cadmium

Cadmium has a harmful effect on both mechanical properties and ease of casting and should not exceed 0.005 per cent, but it is not clear whether or not cadmium promotes intercrystalline corrosion.

Other elements
Care should be taken to exclude other elements such as indium, arsenic, antimony, bismuth and mercury since these again cause intercrystalline corrosion. Manganese and silicon arising from the aluminium used in alloying are sometimes present, but rarely have harmful effect in the concentrations found. The remelting of electroplated scrap may introduce nickel, but not in dangerous proportions.

Mechanical properties
One of the main features of the die-casting process is that the metal, forced under pressure into a cooled die, solidifies almost instantaneously to produce a fine-grained structure. It is necessary, therefore, that properties should be determined on specimens which have themselves been die cast under commercial conditions. Typical values obtained at room temperatures from such test pieces are given in Table 20.

Table 20

Typical properties of pressure die-cast zinc alloys

	BS 1004 A	BS 1004 B
Tensile strength, as cast (MN/m^2)	286	335
Elongation on 5.08 cm, as cast (per cent)	15	9
Hardness (B.H.N.)	83	92
Impact strength (Charpy unnotched 6.35 mm $\times$ 6.35 mm section), as cast (J)	57	58
Solidification shrinkage (cm/m)	1.17	1.17
Thermal conductivity at 20°C $Wm^{-1} K^{-1}$	113	109
Electrical conductivity at 20°C (% I.C.A.S.)	26	26
Thermal expansion (linear per °C)	27×10^{-6}	27×10^{-6}
Density (g/m^3)	6.7	6.7

When using the figures quoted in the table for design purposes, several points should be borne in mind. First, it has already been pointed out that one of the main disadvantages of zinc as an engineering material is that it has no generally recognised modulus of proof stress but tends to flow under continuously applied loads. With their high zinc content the behaviour under stress of the zinc die-casting alloys is also dependent upon time, temperature and strain rate. Under high strain rate conditions, the tensile and impact strength values quoted above are indicative of what may be expected, but under long-term continuous loading, much lower values should be taken. Should the designer wish to use zinc die castings as structural members under continuous loading, *Engineering Properties of Zinc Alloys* [8] gives a range of working stresses and the distortion which they cause after various times of application. This information can be used to guide design. *Designing in Zinc* [9], also published by the

International Lead Zinc Research Organisation, is another useful collection of design information.

A second feature of the alloys which in some cases may be of importance is that they are subject to a small ageing change, owing largely to precipitation in the aluminium-rich phase as is explained later. This results in shrinkage which develops after casting, but this is so small that for most applications it may be ignored. Where the highest precision is required, the castings can be given a stabilising treatment which consists in heating them for 3–6 hours at 100°C, depending on the thickness of the casting, prior to finish machining if this is required.

The effect of ageing on the mechanical properties is also small. After 10 years a reduction of 10 per cent of the tensile strength can occur, but impact strength and elongation are virtually unaffected.

The alloys are notch-sensitive and impact strength values, as determined by the Charpy test, are quoted on unnotched bars; but problems can almost always be overcome by suitable design, avoiding sharp angles and other causes of stress concentration.

The microstructure of the zinc die-casting alloys

As can be seen from Fig. 37 (Chapter 10), aluminium–zinc alloys consist essentially of a zinc-rich phase β and an aluminium-rich phase α, the eutectic containing 5 per cent aluminium melting at 382°C. When molten metal containing 4.0 per cent aluminium, corresponding to the Mazak composition, is cooled, the first phase to solidify is β, containing 0.35 per cent aluminium, and then the remaining liquid freezes as a eutectic consisting of α and β, as can be seen in the micrograph of the as-cast condition, Fig. 51. Immediately after casting, the aluminium-rich phase α begins to precipitate within the β-crystals until their composition reaches 0.07 per cent aluminium, when complete stability is gained — at normal temperatures this takes approximately 5 weeks.

Susceptibility to intercrystalline corrosion

The early zinc die-casting alloys, prepared from relatively impure zinc, suffered heavily in humid atmospheres from intercrystalline corrosion. This differs from orthodox corrosion, in which the attack is general over the whole surface, since it is concentrated and penetrates the metal along the grain boundaries. In the absence of aluminium, zinc does not corrode in this way even when containing significant quantities of other impurities. With an aluminium content in excess of 0.075 per cent, zinc begins to show pronounced susceptibility to intercrystalline corrosion, which is greatly increased if lead, tin, thallium and perhaps cadmium are present in concentrations above 0.01 per cent. The addition of 0.04 per cent magnesium reduces the tendency to intercrystalline attack, and if

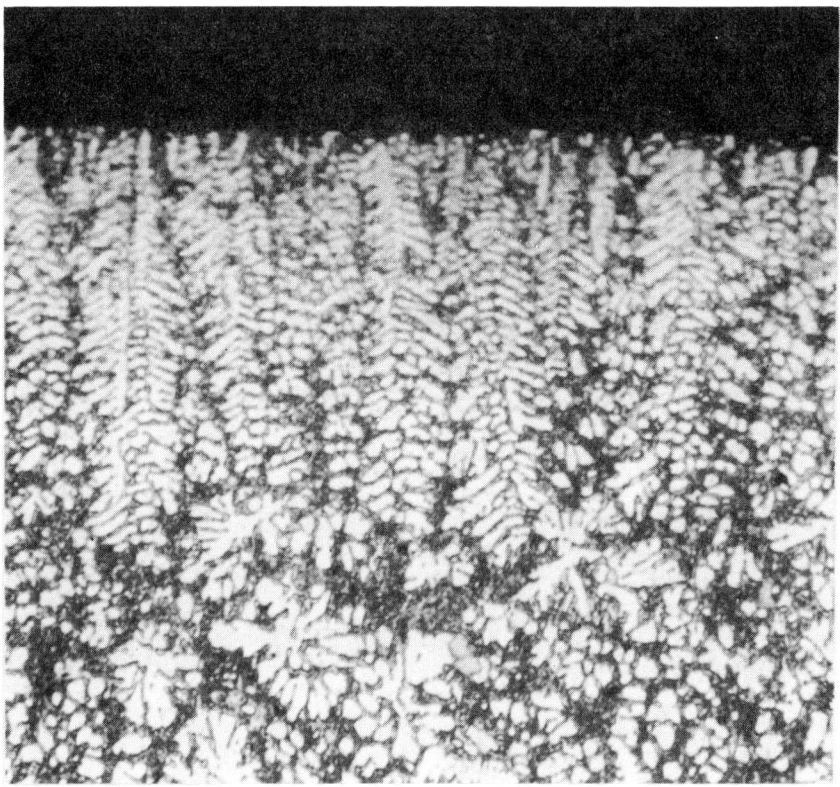

Fig. 51 – Die-cast Mazak 3, as cast. (Magnification ×250.)

the content of impurities is also reduced to the levels laid down in the specification (Table 19), the liability to this type of corrosion is reduced to negligible proportions, and stable alloys can be produced to give good service in a wide range of conditons.

Reasons for the susceptibility of the zinc–aluminium alloys to corrosion of this special type are not clear and the problem is complex. Roberts [10] has suggested that due to the precipitation of the aluminium-rich phase after casting, a concentration of aluminium atoms occurs at the grain boundaries and these may oxidise preferentially. Presumably the deleterious effect of impurities such as lead, tin and cadmium is that they also concentrate at grain boundaries and become poles of a galvanic couple thus accelerating attack in the presence of conducting films. This may well occur but corrosion can take place in dry steam under conditions where aqueous films cannot exist, and electrochemical action is not possible. The protecting action of magnesium has been attributed to the fact that it occurs as the relatively stable compound $MgZn_5$, again at the grain boundaries, and acts as a barrier to penetration.

Temperature effects on mechanical properties
The effect of temperature on the mechanical properties of the BS 1004 Alloys is shown in Fig. 52, the data being supplied by the manufacturers [11].

At elevated temperatures there is some decrease in tensile strength and a more marked decrease in creep-resistance. Zinc die castings, if subject to stress, should not be subject to temperatures above 100°C. One of the features of the alloys is the sudden reduction in impact strength at temperatures below 10°C, and although this drop appears from the curves to be spectacular, it should be remembered that even at −40°C the impact strength of the alloys is greater than that of grey cast iron. However, the loss in impact strength must be taken into consideration in the design of components for service in cold conditions, if shock loading is anticipated.

The die-casting process
One of the advantages of the zinc die-casting alloys is that they can be cast in machines of the submerged-plunger type, since their melting point is relatively low and their attack on steel is only slight. These machines can be operated rapidly with a high degree of automatic control.

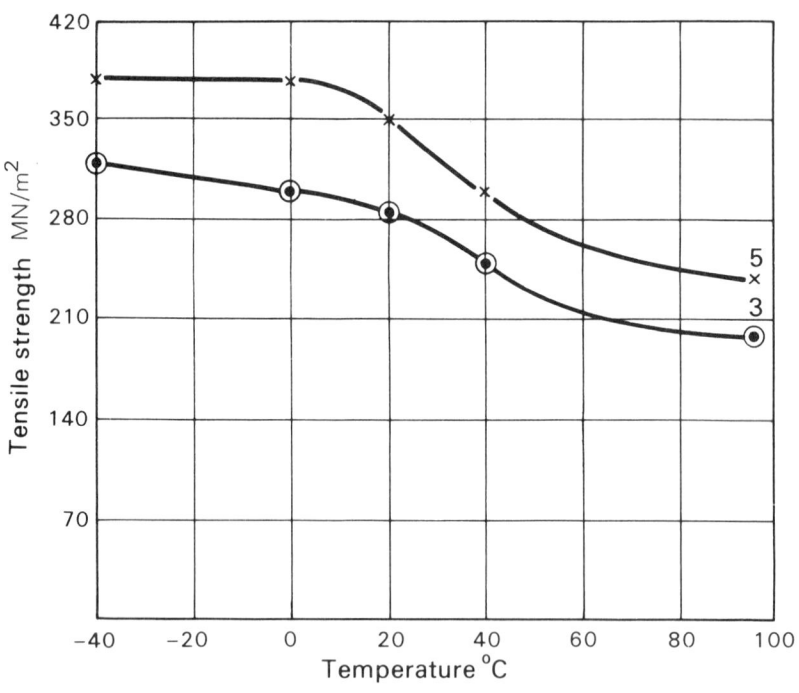

(a)

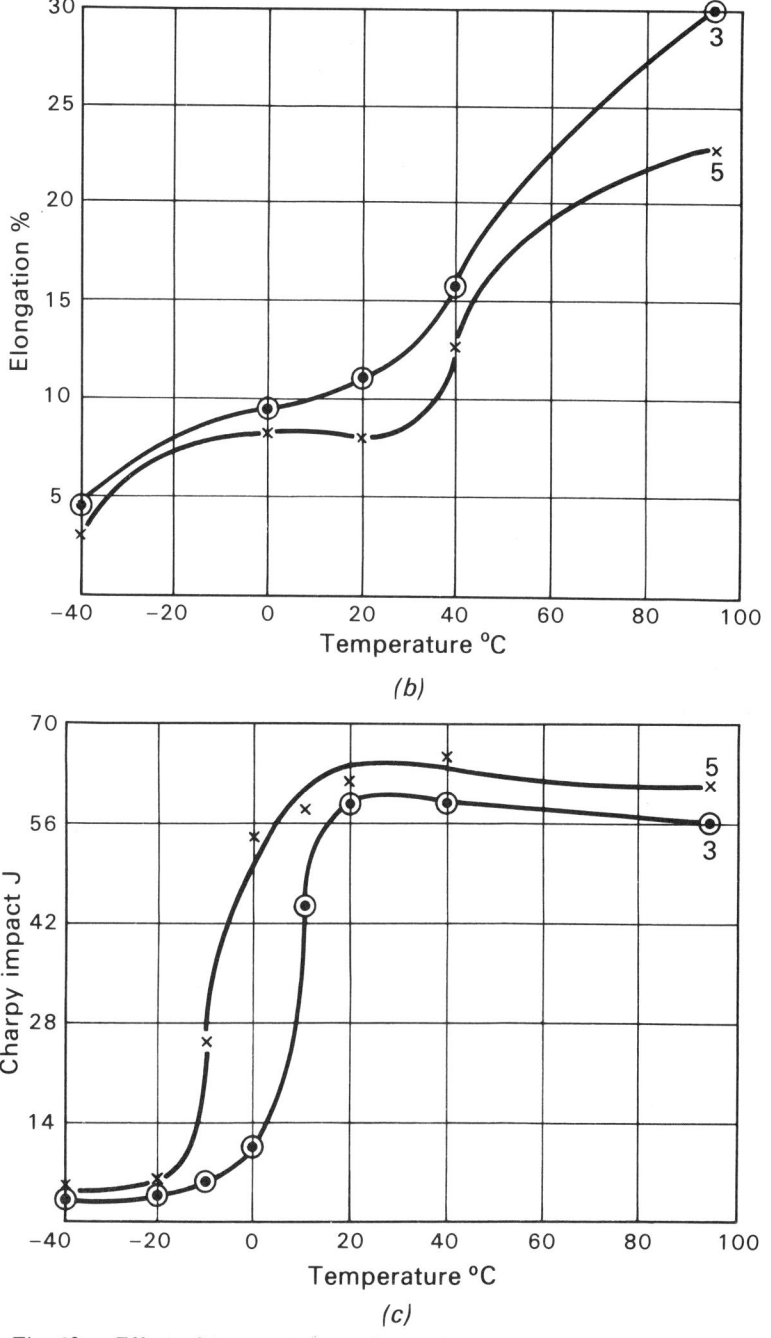

(b)

(c)

Fig. 52 — Effect of temperature on the mechanical properties of zinc-base die-casting alloys. (a) Tensile strength. (b) Elongation. (c) Charpy impact.

The essential construction of such a machine is shown in Fig. 53, which indicates how operation of the plunger forces metal at each shot through the gooseneck and into the die. After a prescribed interval to ensure solidification, depending on the size of the casting, the die blocks are parted and ejector pins force the casting out of the die. The die blocks are then closed, the plunger retracts, the gooseneck refills with metal, and the cycle is repeated. The operation can be sequenced automatically, and a high rate of throughput obtained: for example, on small castings, machines have been operated at 80 shots per minute.

Correct die design is most important and requires skill and experience. Sharp changes of section must be avoided. The gating and venting channels must allow the die to fill as rapidly and evenly as possible and, to avoid surface flow-marks, undue turbulence must be prevented. In recent years the zinc die-casting process has been very thoroughly investigated and a new approach to the design of runners and gates has been developed. The die cavity must be filled fast enough, and with the proper alloy velocity as it passes through the gate, to ensure castings with the optimum soundness and surface finish. Die temperature control is important, as is the need to match the machine's pumping capacity with the die filling requirements [12].

Because solidification is so rapid, some fine internal porosity — of the order or 5–6 per cent — is unavoidable; but it can be positioned to a large extent in unstressed areas, by careful die design. The porosity can thus be well dispersed and zinc die castings rarely need impregnation to make them pressure-tight.

The process is versatile, and a large variety of components are die cast in zinc alloy, some of great intricacy (Fig. 54). The largest weigh several kilograms, and at the other extreme, wedges for wristwatch hairsprings weighing less than 0.1 g are also die cast in zinc alloy.

Their corrosion-resistance against atmospheric conditions and sea water is good, particularly if a chromate treatment is first given, in a weak chromic acid solution, forming a passive film on the surface. They can accept a large variety of paint, lacquer and enamel coatings.

Large numbers of parts are chromium-plated. For such treatment, the castings are usually polished by vibration with abrasive-impregnated plastic cones, to remove traces of parting lines and flow marks, and then plated first with copper, then with nickel and finally with chromium. The usual procedure is as follows [13, 14]: after degreasing and a quick acid etch, plating starts by apllying an enveloping copper layer, as nickel-plating solutions attack zinc alloys, and any zinc contamination so produced affects the quality of the nickel deposit. The initial copper strike must be made from a solution with a low copper ion concentration, to avoid formation of a non-adherent chemically displaced deposit. A copper cyanide solution is usually used for the strike, followed by bright acid copper to build up the deposit quickly to a thickness of 8 μm. This procedure helps to obliterate minor bleminishes in the cast surface.

For castings to be used under mild indoor conditions, a single bright nickel

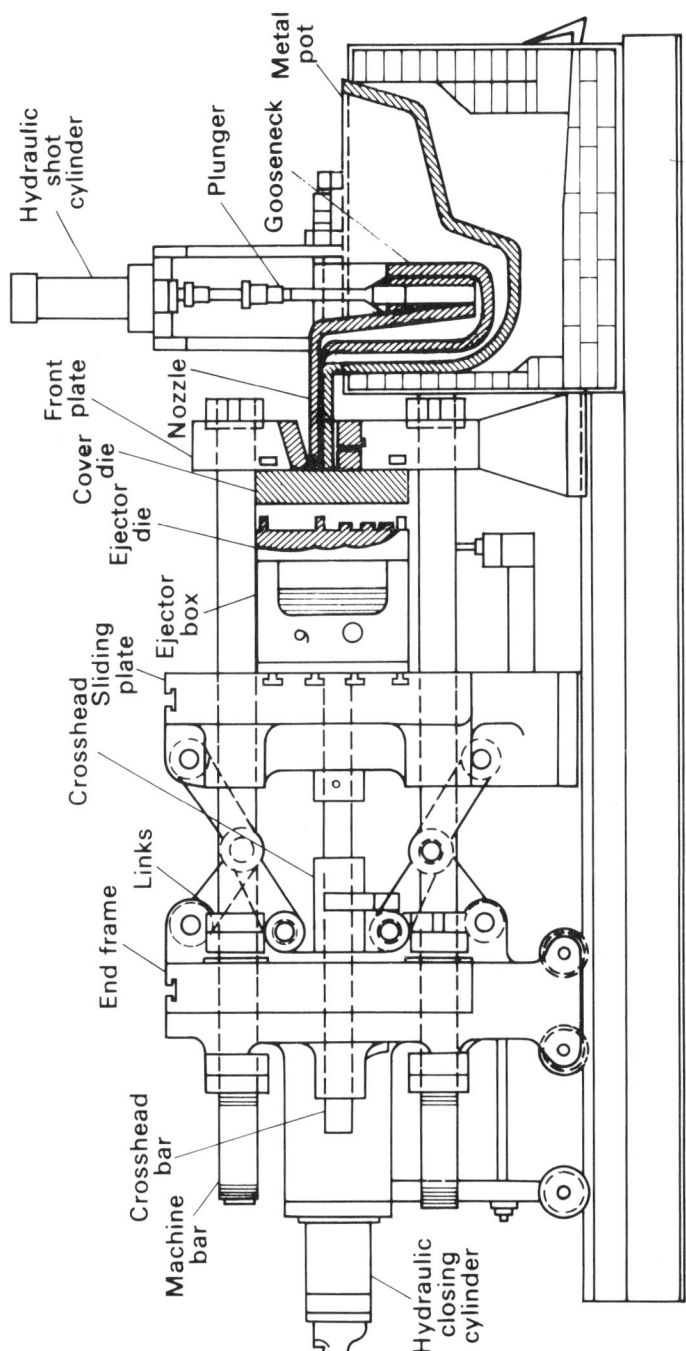

Hydraulic shot cylinder

Plunger

Gooseneck

Metal pot

Front plate

Nozzle

Cover die

Ejector die

Ejector box

Crosshead

Sliding plate

Links

End frame

Crosshead bar

Machine bar

Hydraulic closing cylinder

Fig. 53 – Outline of die-casting machine.

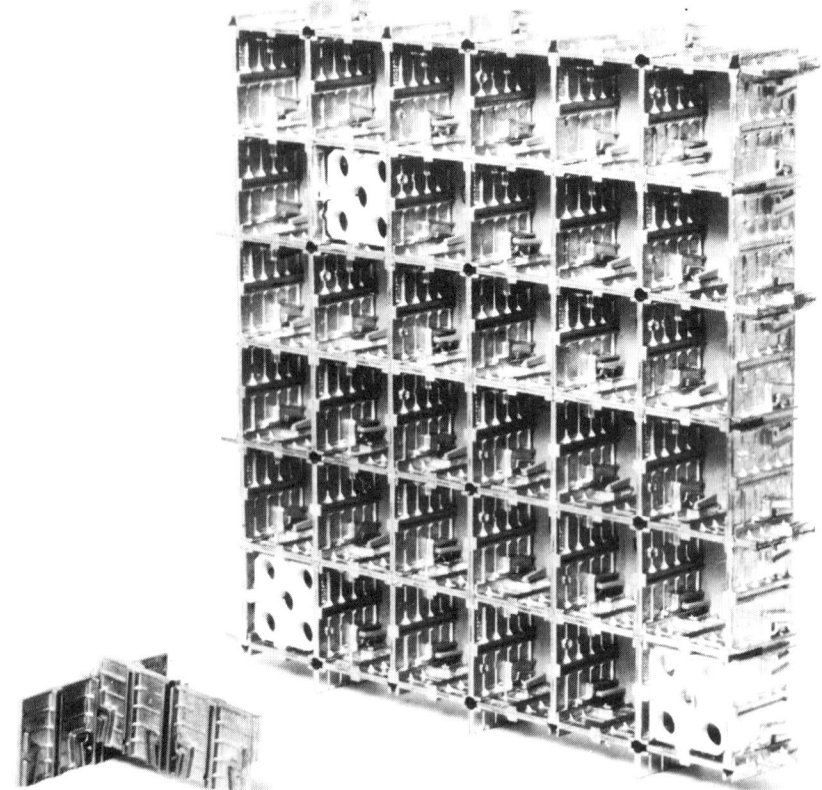

Fig. 54 – Die-casting Mazak 3. Units built up to form a display system to monitor
factory processes.

deposit of 8 μm minimum thickness, followed by bright chromium, is adequate.
For exposure to more corrosive conditions outdoors, a thicker nickel deposit is
required, preferably a duplex system, consisting of a semi-bright nickel deposit,
followed by a thinner layer of bright nickel, with a total thickness of 25 μm.
The outermost chromium deposit is then either microporous or microcracked, so
that any corrosion is spread thinly over the surface, and the light corrosion
products can be easily removed. If conventional bright chromium is used, it
quickly develops a few relatively large cracks, at which intensive corrosion can
lead to blistering and pitting, which can eventually extend into the zinc alloy.

Optimum plating procedures have been laid down in British Standard
Specification BS 1224.

11.5 OTHER ZINC-BASE CASTING ALLOYS

Zinc alloys have been cast for many years. In the nineteenth century, architec-
tural ornaments and garden statues were slush cast using the ordinary crude zinc

then available from horizontal retort manufacture. Its lead content of around
1½ per cent provided a wide enough freezing range for the moulds to be inverted
to let the still molten metal run out after a skin had solidified to make the
hollow casting. Nowadays, pure zinc with about 5 per cent aluminium and
0.04 per cent magnesium finds some applications to make stronger slush-cast
ornaments such as table lamps.

Though zinc alloys with a higher aluminium content and often some copper
— e.g. ALZEN 305, containing 30 per cent aluminium and 5 per cent copper,
developed in Austria during the Second World War as a substitute for phosphor-
bronze in bearings — achieved some commercial success, engineers showed little
real interest in them. During the 1960s the International Lead Zinc Research
Organisation developed an alloy first called ILZRO 12, containing 12 per cent
aluminium and 1 per cent copper with 0.025 per cent magnesium. It was recommended for making sand- or plaster-cast prototypes for zinc die castings as its
tensile strength was similar though its ductility was much less. Eventually, it was
realised that the alloy, after its aluminium content had been reduced to 11 per
cent, was very good for sand or gravity casting, in competition with aluminium
alloys, copper alloys and even cast iron, for numerous applications. Compared
with the other alloys, it is cheaper to melt, and provides better working conditions at the lower casting temperatures. It has the advantage over light alloys
of not constituting a sparking hazard when struck by rusty iron or steel, so it can
be used safely in coal mines and in chemical plant, and it does not need a protective coating like cast iron. The alloy is now known internationally as ZA 12
despite its reduced aluminium content. Its properties are shown in Tables 21
and 22.

Table 21

Physical properties of zinc-base casting alloys

	Alloy 8 gravity cast	Alloy 12 sand cast	Alloy 27 sand cast
Density at 20°C (g/cm^3)	6.30	6.03	5.00
Solidification shrinkage (%)	1.1	1.3	1.3
Solidification temperature range (°C)	404–375	432–377	484–375
Thermal expansion (μm/m °C) at 20°C to 100°C	23.2	24.1	26.0
Thermal conductivity (Wm^{-1} K^{-1}) at 24°C	115	116	125.5
Electrical conductivity (% I.C.A.S.)	27.7	28.3	29.7
Specific heat capacity (J/kg °C) at 24°C to 92°C	435	450	525
Electrical resistivity ($\mu\Omega$ cm) at 20°C	6.2	6.1	5.8
Pattern maker's shrinkage (mm/m)	10.4	13	13

More recently two further alloys have become commercially available, **ZA 27** and **ZA 8**. Their properties are also shown in Tables 21 and 22 and the

Table 22

Mechanical properties of zinc-base casting alloys

	Alloy 8	Alloy 12		Alloy 27		
	cast gravity	sand cast	permanent mould cast	sand cast	[a] sand cast and homogenised	Pressure die cast
Ultimate tensile strength (MNm^{-2})	221–255	275–317	310–345	400–440	310–325	424
Yield strength (MNm^{-2}) (0.2% offset)	208	207	214	365	255	360
Elongation on 51 mm (per cent)	1–2	1–3	1–2	3–6	8–11	1–3
[a] Hardness (B.H.N.)	85–90	92–96	88–90	110–120	90–100	105–120
Shear strength (MNm^{-2})	241	248–262	N.A.	283–297	221–228	N.A.
[a] Impact strength (J) (10 mm × 10 mm bar unnotched) at 20°C	N.A.	27 ± 3	N.A.	48 ± 7	58 ± 12	N.A.
Fatigue strength (MNm^{-2}) (5 × 10^8 cycles)	51.8	103.4	N.A.	172.5	103.5	N.A.
Creep strength (MNm^{-2}) stress to produce creep rate of 0.01%/1000 h at 20°C	59 approx.	59 approx.	59 approx.	69 approx.	86 approx.	N.A.
Creep rate at 138 (MNm^{-2}) stress (% 1000 h) at 20°C	0.2	0.2	0.2	0.1	0.07	N.A.

[a] Homogenised for 3 h at 320°C followed by furnace cooling.   N.A. Not available.

compositions of all three alloys are given in Table 23. ZA 27 is the strongest non-ferrous alloy that can be pressure die cast and has replaced aluminium alloys in applications where they were inadequate. It cannot be cast in a hot-chamber machine as the high aluminium content and casting temperature compared with the 4 per cent aluminium alloys would result in attack on the cast iron goose-neck and piston. The 8 per cent aluminium alloy ZA 8 is best for gravity die casting and is easier to electroplate than ZA 12. ZA 12 and ZA 27 have useful bearing properties which are still being investigated. ZA 12 can also be pressure die cast in a cold-chamber machine.

Table 23

Chemical composition of zinc-base casting alloys

		Alloy 8 (%)	Alloy 12 (%)	Alloy 27 (%)
Aluminium	not less than	8.0	10.5	25.0
	not more than	8.8	11.5	28.0
Copper	not less than	0.8	0.5	2.0
	not more than	1.3	1.25	2.5
Magnesium	not less than	0.015	0.015	0.010
	not more than	0.030	0.030	0.020
Impurities				
Iron	not more than	0.10	0.075	0.10
Lead	not more than	0.004	0.004	0.004
Cadmium	not more than	0.003	0.003	0.003
Tin	not more than	0.002	0.002	0.002
Zinc		remainder	remainder	remainder

11.6 BRASS

The addition of zinc to copper to form brass is the oldest and one of the most important applications of zinc. The art of producing bronze by alloying copper with tin was developed relatively early, and was one of primitive man's greatest technological advances, but, owing to the difficulties inherent in the metallurgy of zinc, brasses do not seem to have been produced until the first or second century BC, when the cementation process was practised in India and China, and also in Europe by the Romans. A mixture of zinc oxide (as calcined calamine) and charcoal was placed in a crucible and covered by a layer of pieces of copper. The crucible was heated to $1000°C$, when the zinc oxide was reduced and formed zinc vapour, which dissolved in, and was retained by the copper. As the zinc content of the copper rose to approximately 30 per cent, the alloy began to melt and ran down to the bottom of the crucible, and at the end of the reaction was cast into moulds—usually made of stone slabs—and later hammered into the shapes required. The process did not involve the separation of zinc as metal.

The method was the main source of brass production for many centuries—it was still practised in this country in 1850. It was not capable of producing alloy containing more than 30 per cent zinc, and, bearing in mind the losses which must have been difficult to avoid, considerable skill must have been required to reach this figure.

When methods for producing zinc were developed in India and China in the thirteenth and fourteenth centuries AD, and the metal became generally available, some brass was made by directly alloying the two metals. This method was more controllable and enabled alloys with a higher zinc content than 30 per cent to be made if required, but the cementation process was presumably cheaper and its use did not die out for many years.

The brasses constitute an important series of alloys, since by varying the composition and heat treatment, a wide range of mechanical properties can be produced with valuable characteristics and excellent corrosion-resistance. They form one of the three major outlets for zinc metal, but as copper is the major constituent of the commercial brasses they are described in detail in the companion book on copper in this series.

11.7 WELDING OF ZINC AND ZINC ALLOY CASTINGS

Zinc sheet and castings can be gas welded, but the operation requires a certain degree of skill and experience. It should be borne in mind that the thermal conductivity and both the specific and latent heats of zinc are low, and thus care is required to avoid overheating. Since a film of oxide forms readily on the molten metal, copious use of a flux is necessary. This is usually made up of equal proportions of zinc chloride and ammonium chloride. There appears to be little experience of arc welding, but resistance, spot and seam welds can be made in sheet up to 3 mm thick, so long as conditions are carefully controlled. Details of the procedure recommended for welding zinc are given a Data Sheet [15].

REFERENCES

[1] Chollet, P., Recent progress in the metallurgy of wrought zinc alloys, *Canadian Metallurgical Quarterly,* Vol. 7. No. 3., p. 177, 1968.

[2] Brown, J. A. and Laurie, G. H., Recent developments in the rapid evaluation of the creep strength of zinc alloys, *Second Conference of Metallurgists,* Canadian Institute of Mining and Metallurgy, Quebec City, September, 1963.

[3] Pelzel, E., *Metallwirtschaft,* Vol. 17, No. 8, p. 788, 1963.

[4] Rennhack, E. H., *Transactions of the American Institute of Mining and Metallurgical Engineers,* Vol. 236, p. 941, October, 1966.

[5]   Rennhack, E. H. and Conard, G. P., *Transactions of the American Institute of Mining and Metallurgical Engineers,* Vol. 236, p. 694, May, 1966.

[6]   Johnson, A. H., *Metallurgical Reviews,* 146, Vol. 15, p. 115, 1970.

[7]   Nicholson, R. B., Engineering outline, *Engineering,* p. 166, 7 March 1969.

[8]   *Engineering Properties of Zinc Alloys,* 2nd Edition, International Lead Zinc Research Organisation, New York, 1981.

[9]   *Designing in Zinc. A Product Development Guide,* First Edition, International Lead Zinc Research Organisation, New York, 1982.

[10] Roberts, C. W., *Metallurgia,* pp. 57–66, August, 1961.

[11] *Mazak Zinc Alloy for Pressure Diecastings: a booklet for Designers,* Mazak, PO Box 181, Bristol BS99, 7AN.

[12] Allsop, D. F. and Kennedy, D., *Pressure Diecasting Part 2 – The Technology of Diecasting and the Die,* Pergamon Press, Oxford, 1983.

[13] Safranek, W. H. and Broomday, E. W., *Finishing and Electroplating Die Cast and Wrought Zinc,* Zinc Institute, New York, 1973.

[14] *Finishes for Zinc Diecasting,* Zinc Development Association, London, 1980.

[15] Data Sheet 48, *Metal Construction,* Vol. 15(6), p. 345, June 1983.

12

Corrosion protection by zinc

The most important commercial application for zinc is protection for steel, when the good corrosion-resistance and the electrochemical characteristics of zinc give it advantages which are almost unique. The methods for protecting ferrous materials with zinc are galvanising, mainly by the hot-dip process but also by electrodeposition; metal spraying; 'sherardising'; the application of zinc—dust paints; and cathodic means.

12.1 PRINCIPLES OF PROTECTING STEEL BY ZINC

Mild steel, because of its excellent physical properties, is one of the main materials used in civil engineering structures, but it has a major disadvantage, in that in most atmospheres it rusts readily, and consequently some form of corrosion protection is almost always required. Such protection must be durable and complete and tailored to the severity of the conditions to be experienced, but to give the necessary protection is frequently expensive – an adequate system can contribute 20 per cent, or more, to the total cost of the completed structure. A comprehensive and authoritative Code of Practice for protective coatings of steel structures, BS 5493, was issued in 1977, and is a good summary of the problems involved.

As shown in BS 5493, zinc coatings applied by galvanising or by metal spraying are generally superior to most other protective methods, and are the only coatings recommended for 'a life to first maintenance' of more than 20 years.

There are two main reasons why zinc coatings are one of the best methods of protecting steelwork against corrosion. The first is that zinc is itself resistant to attack in normal atmospheres – instances have been recorded where zinc roofing sheet has given service for over 100 years. In most atmospheric conditions the rate of attack on zinc is only 3—10 per cent that on steel, and this is due mainly to the grey impermeable film which forms over the surface of the metal exposed to air. This film resists further attack, although in industrial atmospheres containing sulphur dioxide the film is less stable, because it consists largely of basic zinc carbonate which reacts with the sulphurous acid present. Thus, a coating of metallic zinc on steelwork forms a durable barrier affording considerable protection under normal conditions. The second reason arises from the fact that zinc is considerably more electronegative than iron, and when the two metals are in contact in an electrolyte, zinc tends to dissolve, leaving the iron (steel) unattacked.

The usually accepted values for single electrode potentials are given in Table 24. These values are obtained under standard conditions in cells in combination with a hydrogen electrode containing one mole per litre of the ions of the metal in question. Conditions in the corroding films in service and the potential differences which exist under such conditions vary, but there is little doubt that in most types of atmospheric corrosion, zinc is sufficiently anodic towards iron to maintain a strong protective function. This sacrificial type of protection can extend to some extent over adjacent areas, if the zinc coating has been removed locally to expose the underlying steel. Thus, around cracks or other discontinuities in the coating, the zinc corrodes preferentially and the steel is protected. Under such circumstances, zinc corrosion products tend to accumulate in the cracks, giving further protection. Zinc is therefore superior to other coatings applied to steel, such as most paints, plastic coatings and electrodeposited copper, tin or lead which give envelope protection only, and are not effective when the coating is scratched or perforated.

Table 24

Single electrode potentials

Metal	Volts
$Zn^{2+} + 2e = Zn$	−0.761
$Fe^{2+} + 2e = Fe$	−0.44
$Cd^{2+} + 2e = Cd$	−0.401
$Sn^{2+} + 2e = Sn$	−0.136
$Pb^{2+} + 2e = Pb$	−0.122
$Cu^{2+} + 2e = Cu$	+0.344

A further beneficial property of zinc applied by the hot-dip galvanising process is the fact that molten zinc readily attacks iron, forming on the surface a firmly bonded, high zinc-containing film, which can resist abrasion and deformation to a considerable degree.

12.2 THE HOT-DIP GALVANISING PROCESS

Hot-dip galvanising has been practised for over 150 years, and can be used as a batch process (general galvanising) or as a continuous operation. It consists essentially of a pretreatment stage in which the surface of the steel or iron to be galvanised is thoroughly cleaned, followed by immersion in a bath of molten zinc which coats the work with a thin adherent film.

General galvanising
In general galvanising, a wide variety of parts can be treated individually, as they are fed at short intervals through the cleaning stage and then the zinc bath. A considerable degree of mechanisation is employed, and manual handling is reduced to a minimum. The use of large baths over 20 m in length has been developed, so that structures such as columns for railway electrification, television mast sections and complete railway wagons (see Fig. 55) can be treated, but the method is equally applicable to smaller elements such as window frames and steel water tanks. Articles as small as nuts and bolts can also be treated, but with these, the excess zinc whilst still molten is usually removed in a centrifuge. Welding and fastening techniques have been developed which enable sections to be assembled without destroying the efficiency of the zinc coating, so that completely galvanised structures of almost any size can be built up.

Cleaning
The efficiency of the process depends largely on the pretreatment or cleaning of the work prior to dipping, and in order to obtain uniform and rapid attack on the steel or iron surface by the molten zinc, it is essential that all oxide and grease should be removed to permit alloying to take place evenly over all the surface of the steel.

If the condition of the components to be galvanised requires it, they must first be degreased. If the work is heavily scaled or if castings carry adherent moulding sand, it may be necessary to use grit blasting. The articles are then dipped in acid baths to remove oxide films. Hydrochloric acid (approximately 14 per cent by weight) is generally employed in this country, since it can be used cold, but sulphuric acid (10−14 per cent by weight) is also used. Inhibitors are added to the baths to repress the attack of acid on the clean steel surface.

Fluxing
After pickling, the components, now free from scale and oxide films, are rinsed in running water and then treated with zinc ammonium chloride flux $(ZnCl_2 3NH_4Cl)$ to ensure rapid wetting by molten zinc of the now clean surfaces. Two methods of fluxing are used: in the dry method, the work is passed through a tank containing a 30 per cent solution of the flux to which a wetting agent has been added, then dried in an oven and passed directly to the zinc bath; in the wet method, the work is fed immediately after pickling to the galvanising bath, but passes first through a layer of liquid flux floating on molten zinc, to which floating layer of flux, agents such as tallow or sawdust are added to cause frothing which increases the depth of the layer and reduces fuming. The wet method

requires less plant than the dry process, but more fume is evolved, and it is more difficult to prevent flux entrainment on the finished work.

Fig. 55 — A bas-relief replica of a nineteenth-century locomotive. This 2.5 tonne model of a George Stephenson engine (12m long × 5 m high × 0.6 m thick) is being withdrawn from a galvanising bath.

Galvanising

The bath of molten zinc is held at 450–470°C, and since the purity of the zinc is not critical, the less pure grades such as G.O.B. and Prime Western are frequently used. The amount of lead present in these grades, 1.2–1.6 per cent, is beneficial, since as the solubility of lead at the bath temperature is of the order of 1.2 per cent, the excess settles out at the bottom of the bath, protecting it from zinc attack and facilitating dross removal.

The solubility of iron in zinc at this temperature is less than 0.02 per cent, so that the zinc soon becomes saturated with iron and the excess forms a zinc compound known as dross, consisting largely of $FeZn_7$, containing 4–5 per cent iron. This sinks through the zinc but floats as a mush on the lead, and is periodically removed.

Other impurities in the zinc, such as cadmium or copper, play little part in the galvanising process. Additions of tin or antimony (0.01 per cent) are sometimes made, if it is desired to increase the crystal size of the zinc coating, which gives the characteristic pattern known as 'spangle'. Aluminium is sometimes added in small quantities (0.005 per cent) to reduce the formation of oxide (ash) on the surface of the bath, and to improve the brightness of the work. The aluminium content must be closely controlled, since it restricts the formation of the zinc–iron alloy layers, and profoundly affects the thickness and structure of the coating. In continuous strip galvanising described below, it is added for this purpose.

Effect of steel composition on the galvanising process

When mild steel is dipped into molten zinc, the rate of alloy formation depends on the temperature. In silicon-free rimming steels the dissolution of the steel follows a parabolic rate curve from the melting point of zinc up to 485°C. From 485° to 515° the rate of dissolution is linear, and above 515°C it reverts to a parabolic curve. In the parabolic rate ranges, the attack of zinc on iron becomes very slow after a few minutes. In the linear range it goes on continuously. Apart from the difficulty of controlling the coating thickness under these conditions, a steel galvanising pot would quickly be dissolved . When galvanising silicon-free mild steel, the liquid zinc temperature is kept below about 460°C.

In recent years, silicon-killed continuously cast steel has become commonplace. The presence of silicon dissolved in the steel extends the linear range and increases the rate of reaction with molten zinc. The range extension is mainly towards lower temperatures, so that at 1 per cent silicon, it begins at about 430°C and extends to 530°C.

The effect of time of immersion on coating weight is shown in Table 25. Galvanisers of fabricated steel work, such as structural steel, normally quote for a coating weight of 660 g/m^2 minimum and do not wish to exceed this amount significantly. However, if heavy and light sections are dipped together, the light sections reach galvanising temperature before the rest and so can pick up very heavy coatings when silicon steel is used. Even if all the steel is similar in thickness, it may be difficult to limit the dipping time accurately enough. One solution is to galvanise at a temperature in the upper parabolic range, i.e. about 550°C, but this is expensive in energy needed, and a ceramic pot is required, with radiant heaters over part of the surface [1].

Table 25

Effect on coating weight of immersion time in zinc

Si content (%)	Coating weight (g/m²)	
	2 minutes[a]	9 minutes[a]
0.22	200	800
0.22	470	1700
0.42	680	2900

[a]Time of immersion at 450°C.

Certain alloying additions to the zinc nullify the effect of silicon by combining with it to form silicides, and nickel seems the best to use. The recommended amount is 0.10–0.14 per cent nickel in the zinc, added over a period of days as a master alloy. This Technigalva process, developed by Vieille-Montagne SA, is effective for Si contents up to 0.25 per cent. When galvanising steel with a low silicon content, below 0.05 per cent, only 0.08 per cent nickel can be tolerated, otherwise the specified coating thickness cannot be achieved because of the high fluidity of the zinc [2].

Sometimes a steel with a high silicon content is deliberately chosen to fabricate structures on which a thick zinc coating is specified.

Continuous hot-dip galvanising of sheet and wire

For many years, galvanised sheet was made by dipping steel sheets, previously cut to size, individually in a zinc bath, following the general galvanising procedure described. In 1931 T. Sendzimir [3] in Poland pioneered a major step by developing a continuous method of galvanised sheet production, in which cold-reduced steel strip was fed continuously through cleaning facilities and then a zinc bath, to produce what was virtually a new product, with improved properties.

In the Sendzimir process, the strip is uncoiled and passed first through a furnace held at 400–450°C with an oxidising atmosphere, to remove grease and to form a thin coating of oxide over the strip. After this first furnace, it passes through a second, held at a higher temperature, 730–950°C, in which a reducing atmosphere of cracked ammonia is maintained. Under these conditions, oxide films are reduced and the strip is thoroughly annealed. Before the strip leaves this furnace the temperature is reduced to 480–500°C and it enters the zinc bath in a thoroughly clean condition. After such treatment, wetting of the surface and attack by zinc occur immediately. The weight of coating produced is generally controlled as the strip emerges from the bath by using air jets, which can remove excess zinc if light coatings are required, or can hasten solidification, to produce thick coatings. A flow sheet for a typical continuous galvanising bath for steel is shown in Fig. 56.

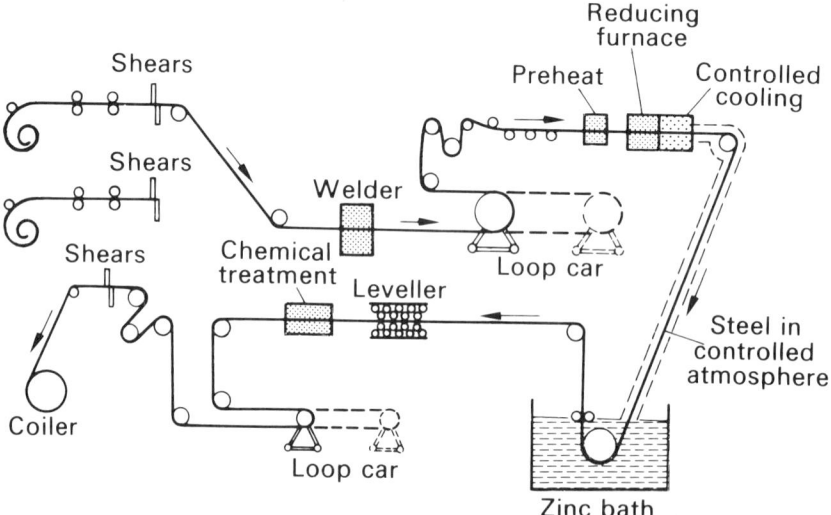

Fig. 56 – Diagram of a typical galvanising line for sheet steel coils.

Variations in practice.

With the widespread adoption of continuous galvanising, a number of variations in the practice proposed by Sendzimir have been developed, mainly in the methods used for the pretreatment of the strip. In the Cook–Norteman process which is widely used in the United States and Canada, the coils of cold-rolled strip are first annealed individually before feeding to the line. Cleaning is carried out by acid pickling and then a thin layer of zinc ammonium chloride flux is applied, which adheres to the strip until it enters the galvanising bath. Other methods of cleaning include an electrolytic etch, used by the United States Steel Corporation. The choice of procedure adopted depends to some extent on the type and quality of the cold-rolled steel to be galvanised.

In the Sendzimir process the galvanising bath can be relatively small and the heat requirement low, since the strip entering the bath is already preheated. The capacity of the line tends to be limited by that of the reduction furnace. In the Cook–Norteman system the capacity of the galvanising bath, which is heated by low frequency induction, is generally the limiting factor. Large capacity lines of both types, treating over 50 tonnes of strip per hour, have been built and coating speeds up to 100 m per minute are reached.

The development of the continuous galvanising line opened up a new dimension in the already considerable market for galvanised steel sheet. Owing to the nature of the operation, labour costs are low, and the process can be closely controlled, either automatically or manually, and a uniform product can be consistently maintained. However, the outstanding advance made by the process lay in the nature of the coating produced.

The coating

A distinctive feature of all continuous galvanising lines is the use of a relatively high (0.1–0.2 per cent) addition of aluminium to the bath. As will be described later, aluminium at these concentrations has a marked effect on the mechanisms of the attack of zinc on iron, suppressing almost entirely the formation of brittle iron–zinc compounds in the coatings produced.

As will be seen from the micrograph, Fig. 57, the typical structure of continuously galvanised sheet shows a thin layer consisting largely of Fe_2Al_5 covered with a layer of pure zinc. The total thickness of the coating is usually about 25 μm, but thicker or thinner grades are available. Such a coating is considerably more flexible and able to withstand greater deformation than the coatings which contain the iron–zinc compounds produced in aluminium-free galvanising baths. They give coatings which tend to be thicker – up to 50 μm on thin sections and up to 150 μm on thick steel. A thinly coated continuously galvanised steel sheet can be bent back on itself without fracture of the coating.

Fig. 57 – Micrograph of continuous galvanised sheet, showing absence of alloy layers. (Magnification ×200.) (Courtesy of BNF Metals Technology Centre.)

With this flexibility such coatings can withstand drawing and forming operations, thus extending the number of applications for which such sheet can be used.

Development of alloy coatings

Many years ago it was realised that the iron/zinc alloy layers present in products galvanised after manufacture had, in addition to excellent abrasion-resistance, better resistance to corrosion than pure zinc, in industrial and acid environments. More recently, work has shown that up to four times the corrosion-resistance of zinc can be obtained by alloying it with aluminium. The first major commercial application was in Galvalume, patented by the Bethlehem Steel Corporation, which is a 55 per cent aluminium, 43.5 per cent zinc and 1.5 per cent silicon alloy, and is used for the continuous processing of steel strip. As well as at Bethlehem Steel, there are now additional lines in the USA, Australia and Europe. The high corrosion-resistance is, however, achieved at the expense of lesser sacrificial protection at edges and scratches, whilst unattractive black staining may appear in certain circumstances. There are also some of the same limitations as aluminium has with regard both to resistance to attack in concrete and similar alkaline media, and the different joining techniques compared with zinc.

The interest generated by Galvalume encouraged development of an alloy with many of its advantages but without some of its technical and processing disadvantages. A 5 per cent aluminium alloy (with around 0.1 per cent mischmetal additions) has now been developed by ILZRO and is known as Galfan. This gives a substantial increase in corrosion-resistance compared with pure zinc but retains good sacrificial protection at edges and scratches. Galfan-coated steel can, in general, be formed and fabricated in the same way as steel with a traditional galvanised coating; indeed tests show it to have still better formability than conventional galvanised steel. The mischmetal improves the appearance of the coating and eliminates bare spots and uneven thickness. The coating requires zinc of minimum 99.99 per cent purity and aluminium of minimum 99.9 per cent purity. An additional advantage of Galfan is that it can be applied using Sendzimir-type plant with a new cast iron pot to avoid contamination of the alloy. The bath operates at about $430°C$ compared with $600°C$ for Galvalume, which requires a refractory pot.

The Galvalume coating shows no preferred crystal orientation of the (III) plane parallel to the sheet surface, whereas conventional galvanised strip and Galfan strip produced with minimum spangle by rapid quenching show a strong tendency for the basal plane to be orientated nearly parallel to the surface. The Galfan coating with normal spangle has the basal plane orientated at up to $50°C$ to the surface with smaller grains that are hexagonal compared with the larger irregular grains in conventional galvanised strip.

So far this alloy has only been adopted commercially by one wire galvaniser in France, but a consortium of steel companies throughout the world (including British Steel Corporation) have invested money to develop the process further.

In test runs, many thousands of tonnes of steel product should be in regular commercial production from 1984.

One-side galvanising

Car manufacturers have found it necessary to use zinc coatings on critical areas of bodies to avoid the early rusting that appeared when vast quantities of salt mixed with grit began to be used on North American roads, to keep them clear in the winter. Normal galvanised strip could be used, but would not provide the desired high quality finish on the outside when painted. Zinc-rich paints were used to protect the inner surfaces and a paint containing zinc dust and zinc chromate achieved widespread use in Europe as well as North America. Ways were also developed of galvanising steel strip on one side only. The simplest involves keeping the strip taut as it passes under a coated steel roller partially immersed in the molten zinc, so that the zinc cannot contact the upper surface.

Other ways of producing strip with only one side zinc-coated involve removing the coating from one side of normal galvanised strip by grit blasting, or by electroplating zinc on to the underside of strip as it passes just below the surface of the electrolyte. However, the former process is expensive and the latter not really economic to provide coatings of the necessary thickness.

Galvanising of wire and pipe

An important application of the hot-dip galvanising process is in the production of steel wire, which is widely used in agriculture and other outdoor applications. Selection is made of the type of steel used for the wire, silicon-killed steels containing approximately 0.07 per cent silicon being most frequently employed, since these give good adherence of the zinc coating. In the galvanising operation coils of wire of the required gauge are fed first through one or two baths of molten lead held at 650–700°C to anneal and soften the wire and to burn off combustible matter. The wire is then passed through a pickling bath containing hydrochloric acid, followed by a rinse in a bath containing dilute ammonium chloride solution, although the gaseous reduction as developed by Sendzimir for sheet galvanising is now used to an increasing degree. The wire, now thoroughly clean, enters the galvanising bath through a flux layer of zinc ammonium chloride and is immersed for several seconds, the period being determined by the thickness of coating required. The temperature of the galvanising bath is held between 450 and 480°C. Emerging from the bath the wire passes through a layer of oil and charcoal, and some of the excess zinc is removed from the still-molten surface. By varying the bath temperature and immersion time, coating thicknesses varying from 7 μm to 20 μm can be obtained.

Steel pipe is galvanised both internally and externally in equipment similar to that employed in general galvanising. The pipe is generally cut into lengths 6.5 m long and treated in batches, chain mechanisms being widely used for handling in and out of the baths. After it emerges from the galvanising bath,

excess zinc is removed from the outside using a blast of superheated air and
steam on the inside surfaces.

The bonding of zinc to iron

The hot-dip galvanising process depends upon the ease with which molten zinc
attacks iron at galvanising temperatures of 450–470°C to form a number of
compounds. The zinc end of the zinc–iron equilibrium diagram, Fig. 58, shows
the components which can form at various temperatures and concentrations
under equilibrium conditions.

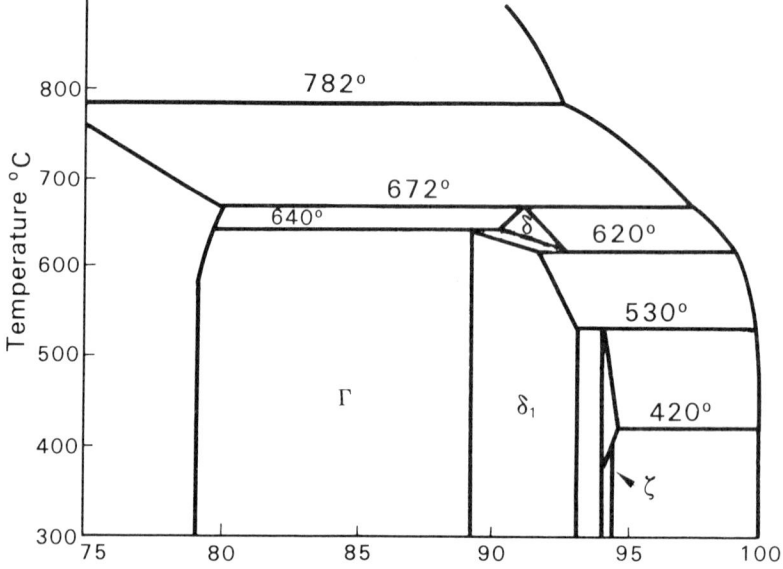

Fig. 58 – The zinc end of the zinc–iron equilibrium diagram.

When steel is immersed in the galvanising bath, zinc atoms begin immediately
to diffuse inwards into the iron lattice, and iron atoms to move outwards. Owing
to the difference in melting points the mobility of the zinc atoms is greater than
that of iron. A series of alloy layers begins to form as indicated diagrammatically
in Fig. 59.

Of the three main phases of interest, gamma, containing 72–79 per cent
zinc, is composed of three compounds – $FeZn_2$ at 83.1 per cent zinc, Fe_3Zn_{10}
at 79.6 per cent zinc, and $FeZn_3$ at 77.9 per cent. The gamma crystallises in the
cubic system, and delta, 87–93 per cent zinc, is hexagonal, containing $FeZn_7$ at
89.2 per cent zinc and $FeZn_{10}$ at 92.1 per cent. Zeta, 94 per cent zinc, is mono-
clinic. A micrograph of a typical hot-dipped galvanised sheet is shown in Fig. 60.

Since in a galvanising bath the time of immersion of the steel is relatively
short, equilibrium conditions are not reached, and all the layers predicted by the

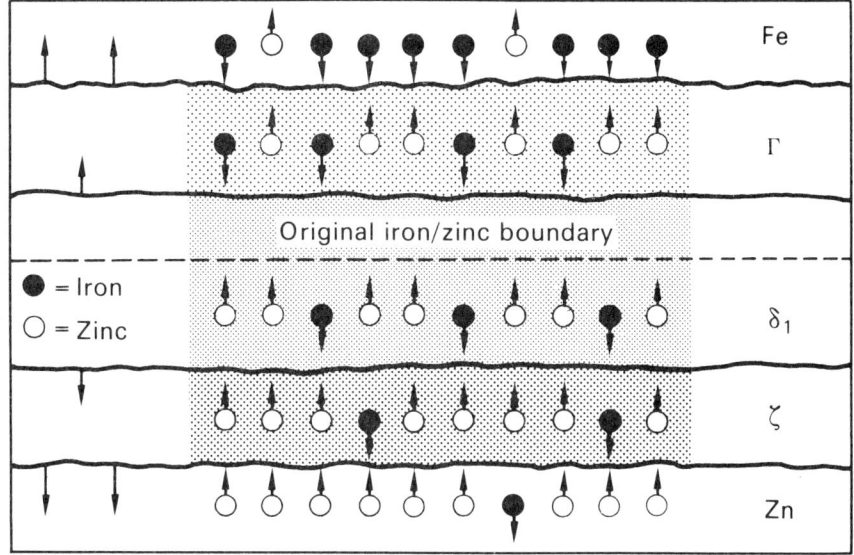

Fig. 59 – Diagrammatic representation of the diffusion of iron and zinc through the alloy layers.

diagram are not always observed. The gamma layer is always thin, and rarely appears in micrographs as a distinguishable phase. It has been shown that the zeta layer forms first and tends to grow fastest, generally becoming the main

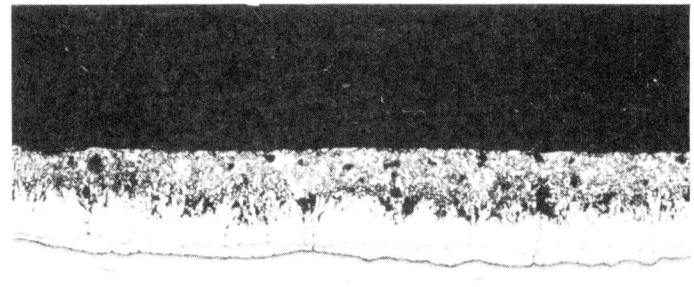

Fig. 60 – Micrograph of hot-dipped galvanised sheet. Galvanised for 2 minutes at 450°C. (Magnification ×200.) (Courtesy of BNF Metals Technology Centre.)

component of the coating. The delta layer can frequently be subdivided into a fine-grained coherent layer next to the gamma, and a palisade-like section, the columns running at right angles to the steel surface. The alloy layer as a whole is obviously under considerable stress since warping can occur if thin sheet is galvanised on one side only.

Rate of attack of zinc on iron
The rate of attack by zinc on a low carbon steel between the range 440–490°C is parabolic with time, as given by the formula

$$L = n\sqrt{t}$$

where L is the weight loss per unit area, t is the time and n is a constant.

At 500°C the rate of attack becomes linear, the loss in weight then being directly proportional to the time of immersion, but above 530°C the rate reverts again to the parabolic form.

In the temperature range used in commercial galvanising, 450–470°C, the alloy layers are compact and relatively uniform, and growth is determined by the rate of diffusion of zinc and iron through the layers, following the parabolic relationship. The change to linear attack at 490°C is generally attributed to incipient breakdown of the zeta phase allowing direct contact of zinc with the layers below. Whilst below 490°C the overall thickness grows parabolically, the relative thickness of the individual layers changes with time at a more linear rate, the delta layer tending to grow at the expense of zeta. The thickness of the outer layer of zinc depends mainly on the withdrawal rate and is independent of the time of immersion.

Effect of aluminium in the bath
It has already been pointed out that aluminium additions have a marked effect on the galvanising reaction. In general galvanising, additions of less than 0.005 per cent aluminium are sometimes made to reduce surface oxidation of the molten zinc and increase the reflectivity of the surface, but at this level little effect on the type of coating can be seen. At the higher concentrations used in conventional continuous galvanising practice, 0.2–0.3 per cent aluminium inhibits entirely the formation of iron–zinc compounds, and forms instead a thin alloy layer on the iron, which has been identified as Fe_2Al_5. This layer grows slowly and during the short time of immersion in continuous baths protects the steel almost completely from further attack.

Galvanising practice
Detailed descriptions of the galvanising process and of the considerable amount of investigational work which has been carried out on the process (much of it at the BNF Metals Technology Centre, Wantage) are given in the references [3]–[6].

Specifications
BS 2989:1982, Hot dip zinc-coated steel sheet and coil, covers the properties and characteristics of the main types of galvanised steel sheet produced. BS 5493:1977 is a general Code of Practice for protective coatings of steel structures, and gives a useful comparison of the values of zinc coatings with other protective systems. BS 729:1971 covers general galvanising.

Coating weights
For steel of a particular thickness, the weight of the coating produced during galvanising depends primarily on the composition of the steel, and the roughness of its surface. Process variables such as immersion time and bath temperatures influence the thickness produced, to a lesser degree. As the thickness of the steel increases, so does the weight of zinc taken up. BS 729:1971 recommends a minimum coating of $610 \, g/m^2$ for material more than 5 mm thick, and $335 \, g/m^2$ on material less than 2 mm thick. In certain cases BS 729 permits the specification of higher or lower coating weights. If required, higher coating weights over $900 \, g/m^2$ can be obtained by using a steel with 0.3 per cent silicon, or by grit blasting the work before galvanising.

Provided the part to be galvanised has been so designed that, on immersion, both pickling liquor and molten zinc can reach all sections, both internal and external, then, since the attack of zinc on the steel is an alloying one, the whole surface of the work will be adequately coated. Coating formation is, to a large degree, self-regulating and the standard BS 729 minimum coating weights can be achieved without difficulty.

Most ASTM specifications are similar to those of BS 729. Most continental (European) specifications, and the ISO standards require lower coating weights than those of the equivalent British Standards.

In the latest Code of Practice, BS 5493, the coatings are described in terms of thickness of coating rather than coating weight. The relation between the two systems can be calculated from the equation $304 \, g/m^2 = 1 \, oz/ft^2 = 43 \, \mu m = 0.0017 \, inch$.

Tests for coating weight
Portable magnetically operated instruments available, such as the BSA Tinseley Pencil Gauge or the Elcometer Inspection Gauge, give rough guides to coating thickness. Such tests are non-destructive. An accurate determination of coating thickness can be obtained only by stripping tests, or by direct measurement of typical micro-sections.

Design for galvanising
When galvanising is to be the chosen method of protection to be applied, then the characteristics of the process must be fully considered at the initial design stage. The process relies on the rapid attack of molten zinc on steel. Since the

zinc must thoroughly 'wet' the steel, the surface of the latter must be thoroughly cleaned during the pickling treatment. It is thus essential to provide ready access on immersion, to all surfaces, to both the pickling liquors and subsequently the molten zinc, and also to ensure that on withdrawal, adequate drainage occurs. No closed or partially closed sections, where access or drainage is impeded, should be permitted. Sufficient drainage and venting holes must be provided. Internal and external stiffeners, baffles, gussets, diaphragms, etc. should have corners cropped to aid the free flow of molten metal.

The question of design and other aspects of the galvanising process are discussed in a useful brochure published by the Galvanisers Association [6].

Welding of galvanised steelwork
Welding can be carried out either before, or after, galvanising. If carried out before, care must be taken to see that the welds are free from porosity or lap joints, which could trap acid during the pickling stage prior to galvanising. This could exude later, causing staining.

Galvanised steel can be joined by the same methods as used for uncoated steel, and the presence of zinc has little effect on the properties of the welded joints. All forms of resistance, arc and also oxyacetylene welding can be used, although resistance welding is usually restricted to sheets less than 3 mm thick. If after welding, areas devoid of zinc are left around the weld, they must be protected, either by zinc spraying or by an application of zinc-dust paint.

When galvanised steel is fusion welded, it is difficult to avoid the evolution of fumes of zinc oxide. These can be copious and unpleasant, and if inhaled in sufficient quantity can cause zinc fume fever. Whilst this can lead to feelings of nausea, fits of shivering, and in extreme cases vomiting, the effect is not lasting, and complete recovery occurs within hours. The effect can generally be avoided entirely by adequate ventialtion of the welding operation.

A summary of the techniques involved in the welding of galvanised steel is given by F. C. Porter [7].

12.3 ZINC COATING OF SHEET AND WIRE BY ELECTRODEPOSITION

Most zinc-coated sheet and wire is produced on continuous hot-dip galvanising lines, but some is galvanised electrolytically, since for a number of applications the electroplated coating is superior. Although the adherence and ductility of the hot-dipped coating is good, the electrodeposited zinc coating is even better and can withstand severe forming operations. In addition, the smoothness and appearance of electroplated zinc is superior. The coating thickness of electro-galvanised sheets is low, varying from 2.5 to 4.5 μm, whereas the thickness of zinc on contiuous galvanised sheets is generally of the order of 25 μm, giving considerably greater resistance to atmospheric attack. Wire can be plated with zinc up to 75 μm thick for subsequent drawing to leave a coating about 12 μm thick.

Strip to be plated is always in the cold rolled condition. It is fed continuously through cleaning baths and then immediately into the plating bath. The electrolyte used is basically a solution of zinc sulphate or chloride, although other salts are added to improve conductivity. Anodes of high purity zinc 99.99 per cent pure, and better are used, and direct current is supplied at approximately 12 volts to give current densities varying from 1300–2700 A/m^2. The temperature of the bath is kept below 65°C. Usually, one surface of the moving strip is coated at a time from below, the direction of travel then being reversed over rolls and the strip passing in the opposite direction so that the previously unplated surface is coated in its turn. After plating, the strip is rinsed, and is generally passed through a bath of free phosphoric acid to give a thin film of zinc phosphate which forms a good basis for paint or other finishes. After the phosphoric acid treatment, a final rinse is given in dilute solution of chromic acid which passivates the zinc surface and affords some protection during subsequent handling and storage. For some applications the phosphoric acid treatment is omitted and only the final chromic acid rinse used.

Zinc plating of fabricated items

Zinc plating is used to protect small hardware items such as nuts, bolts and wood screws. After degreasing and pickling, they are loaded into perforated barrels made of insulating materials but containing a cathode contact. The barrels are rotated in a tank containing a solution of a zinc salt − often zinc cyanide with sodium cyanide − and a zinc anode. As the barrel rotates, the parts to be plated tumble around in contact with the cathode and so get uniformly plated. The anode zinc dissolves and has to be replaced. It is often added as gravity cast pure zinc balls retained in a basket.

Even when bright zinc plating solutions are used − they contain organic compounds which control the crystal structure of the deposit − the parts often need a dip in 1 per cent nitric acid to brighten their surface. Subsequent treatment with an acidified chromate solution followed by an alkaline rinse helps to preserve the attractive finish. Coatings are usually in the range 3–9 μm thick according to the service requirements.

Larger parts can be zinc plated, and high tensile steel fasteners are often protected in this way. Steel that has been severely worked or that has a tensile strength greater than about 1000 MNm^{-2} should be stress-relieved before plating. Such items also need heat treatment immediately after plating to minimise the embrittlement caused by the migration of hydrogen atoms into the steel during plating.

Bright zinc plating can be applied to zinc die castings to provide a cheap bright finish.

12.4 SPRAYED ZINC COATINGS

Whilst hot-dip galvanising is the method in most general use for the cathodic protection of steel against corrosion, it cannot be used in a number of applications, particularly where the work must be treated in situ. For such uses, sprayed zinc coatings offer a good alternative.

The process of metal spraying was invented in 1910 by a Swiss engineer, Dr Schoop. The method consists essentially in feeding zinc wire, up to 5 mm in diameter, through an electric arc, or an acetylene or propane flame, in which the zinc melts and is broken up by turbulence into fine molten droplets. These are forced at high velocity on to the work to be treated. On impact at the steel surface, the droplets solidify, and adhere tenaciously both to the steel and to each other, forming a coating with considerable mechanical strength. The coating is porous (5–10 per cent volume), since each of the droplets is surrounded by a thin film of oxide. In the case of zinc, its anodic properties ensure that the protection given is comparable to that of a hot-dipped coating of equal thickness. The arc pistols are the most economical to use, and can spray up to 50 kg of zinc per hour.

In the case of zinc, an alternative method feeds the metal to the flame as high grade zinc dust. The coatings produced by this method are similar in properties to those from wire fed pistols, but the latter is more flexible and more widely used. The capacity of the pistols varies — an average value for a flame pistol is 30 kg of zinc sprayed per hour.

In a number of conditions, in rural or mild urban atmospheres, both zinc and aluminium sprayed coatings give good service. In more aggressive situations such as marine atmospheres, sea water immersion, and industrial areas, sealed zinc coatings are preferable. Sprayed zinc coatings are superior for use in alkaline waters, particularly where the carbonate content is high. In contact with hot water, aluminium coatings are probably preferable. The effect of various types of supply waters on zinc and aluminium coatings is discussed by Campbell [8]. The questions involved in the choice of protective systems for structural steelwork are covered by BS 5493.

For many spraying applications, a zinc — 15 per cent aluminium alloy has been found more suitable than pure zinc, and at this composition (and indeed in compositions up to about 18 per cent aluminium) it is readily produced in the form of wire. Special alloys can also be applied by metal spraying, and the zinc/copper/titanium alloy (developed as a creep-resistant roofing alloy), when sprayed on the fraying or contact surfaces of joints, gives a reliable combination of corrosion protection and creep-resistance. Traditionally such joints have been made with the steel left unprotected, to provide full frictional characteristics, but this left a problem of protecting the joints from corrosion.

Before spraying, the steel must be clean and free from scale: this is usually attained by grit blasting. The thickness of coating applied varies from 50–250 μm depending upon the protection required, which is proportional to coating thickness. If severe conditions are to be encountered, organic fillers, such as vinyls or chlorinated rubbers, are applied to seal the pores in the coatings. Sprayed zinc coatings form a good basis for a wide variety of paints, and the combination of a thin sprayed coating and a suitable paint offers excellent protection under most conditions. Figure 61 shows the Tamar Bridge which was sprayed prior to painting.

Fig. 61 – Bridge over the River Tamar. The bridge was zinc sprayed and then painted. (Courtesy of the Zinc Development Association.)

12.5 SHERARDISING

In 1900 Sherard Cowper-Coles, an inventor of considerable ingenuity, developed a process for coating small steel articles with zinc as an alternative to hot-dip galvanising. The process, which bears his name, is still in use today, for coating nuts and bolts, small castings and sections of pipe or conduit.

The process is carried out in rotating steel drums which are heated either electrically or by gas to 370–429°C. The steel parts to be treated are cleaned in an acid pickling bath, rinsed and dried, and then packed in the drum with a mixture of metallic zinc dust and sand. The drum is not completely filled so that some movement can occur from the tumbling action as the drum is rotated. The drum is then heated to the required temperature and rotation maintained for 6–10 hours depending upon the coating thickness required.

As treatment proceeds, diffusion of zinc into the steel surface occurs initially to form a zinc—iron bond. Zinc then builds up on the surface and a firmly adherent coating, resistant to both bending and mechanical shock, is obtained. If the articles are of irregular shape, without re-entrant angles, uniform coatings can be obtained. Depending on the time and temperature of treatment, the thickness of coating developed can vary from 0.015 to 0.025 mm. Corrosion-resistance is comparable to that of other zinc coatings of equivalent thickness.

Mechanical plating

This cold, non-electrical process is also known as peen plating. The degreased steel parts are given a very thin coating of copper, generally by dipping into a copper solution, and are then tumbled in a barrel with a calculated quantity of zinc dust, water, small glass beads (ballotini) and a wetting agent. After about an hour a fairly bright zinc coating is formed by cold welding under impact from the glass beads. The thickness is around $10-25$ μm. The main advantages of the process compared with zinc plating or sherardising are the avoidance of hydrogen formation and of high temperatures, so it is used mainly to protect high tensile or hardened steel, e.g. springs.

12.6 ZINC-DUST PAINT

The metallic content of most commercial grades of zinc dust is in the range $95-97$ per cent. The average particle size of standard grade dust lies between 5 and 9 μm, but dust with an average particle size of only 2.5 μm is made for certain applications. The particles are mainly spherical in shape, and covered with a thin, firmly adherent film of zinc oxide.

In the first applications of zinc dust to paint formulations a ratio of 80 per cent zinc dust to 20 per cent zinc oxide was used. These were first used as primers for galvanised iron, but it was soon found that such paints could give protection to ungalvanised steel surfaces and their use for this purpose expanded. A considerable advance was made at Cambridge by Britton and Evans [9] and Evans and Mayne [10] in the period 1942–3, when it was shown to be advantageous to use high concentrations of zinc dust (up to $90-94$ per cent by weight) in the film. Such films have sufficient electrical conductivity to give a degree of the sacrificial protection afforded by galvanised or sprayed zinc coatings.

Many formulations are now available of good quality zinc-rich paint using a variety of organic and inorganic vehicles, and good protection can be given in most atmospheres to steel surfaces, if they have been cleaned – usually by shot or sand blasting. The organic binders used are of three main types – chlorinated rubber, epoxy-polyamide resins and phenoxyresins. The paint is usually applied by airless spray, but most other methods of application can be used. For good protection the thickness of the dried film should be in the range $60-125$ μm. The inorganic binders used are mainly sodium, lithium or ethyl silicates. Maximum hardness and adhesion is sometimes developed by baking the coating at 200–230°C for a short period after application, but a number of formulations are self-curing. The silicate-based paints are particularly useful in marine environments.

Zinc-dust paints can give excellent service, but the duration of protection is limited, depending primarily on the adherence of the vehicle, and galvanising or zinc spraying is more permanent. However, zinc-dust paint offers one of the best alternative methods of protection if the work cannot be galvanised.

12.7 ZINC ANODES FOR CORROSION PROTECTION

The rusting or corrosion of iron is electrochemical in character. In practice, a steel surface is rarely uniform, and slight differences in composition, grain size and lattice strain, due to variations in degree of cold work, cannot be avoided, causing slight variations in the electrochemical potential of the steel from area to area. Under a film of moisture, which acts as an electrolyte, cells are set up and corrosion of the more anodic areas begins.

That zinc can be used in such methods as galvanising, electroplating or spraying to protect steelwork is due to this effect, since in most corroding media in practice, zinc is considerably more electronegative than iron. By dissolving or corroding preferentially, the zinc thus protects the areas of iron adjacent to it.

This principle is applied to protect from corrosion large areas of steelwork which cannot be galvanised, such as ships' hulls (see Fig. 62) or steel pipe lines.

Fig. 62 – Zinc anode protection – typical arrangement around a ship's stern.

Plates of zinc are attached to the steel at suitable intervals. As the steel corrodes, the zinc becomes anodic, rendering the surrounding areas cathodic and greatly reducing corrosion. The difference between the electrode potential of zinc and iron is sufficient to negate the small local differences over the steel surface which form the major cause of normal attack, and long life can be achieved.

Similar superficial protection can be provided by any metal more electro-negative than iron, and aluminium and magnesium are also used to give cathodic protection. However, magnesium, though offering excellent electrochemical protection, tends to corrode rapidly, whilst aluminium anodes used in sea water can become passive, and inactive, under certain conditions. Although zinc can also suffer similarly owing to the build-up of impervious films of corrosion products, this can be largely avoided when metal with a low iron content (less than 0.005 per cent) is used.

For an anode to be fully effective, it is essential that it remains in electrical contact with the protected structure throughout its life. In order to ensure this, the zinc is generally cast around galvanised steel straps, which protrude from the zinc, and these can then be welded to the structure. When suitably installed, under normal conditions, the efficiency of the zinc is high, and over 90 per cent is usefully consumed, protecting the adjacent steel. Depending upon the conditions, such anodes can be expected to have a useful life of up to 25 years, and if they can then be replaced during dry docking, or by divers, a high degree of protection can be given almost indefinitely. Care must be taken to see that the anode surfaces are not painted.

Zinc anodes have proved particularly successful in protecting pipes carrying oil or gas, lying on the bed of the North Sea. They are also used in the holds of oil tankers to give protection when the holds are ballasted with sea water. For this purpose, the non-sparking property of zinc is an additional safety factor.

When using zinc anodes to protect steel structures such as ships hulls, ballast tanks, buried or immersed pipe lines, many factors must be taken into consideration in deciding the number and siting of the anodes used. The area to be protected, the corrosive character of the soil or water in which the structure is to be immersed, and the presence of strong currents are but some of the factors on which the efficiency of the protection to be given can be based. While zinc anodes are not effective in all circumstances, there are many applications where they give essential protection for a long time to fabricated steelwork which cannot readily be renewed, and their use in the marine and oil industries is growing.

12.8 ECONOMICS OF PROTECTION

As has been stated, because of the readiness with which iron rusts when exposed to the atmosphere, most steelwork requires some protection. The choice is generally between paint or plastic systems, and zinc coatings, applied either by galvanising or spraying. Initially, paint systems are somewhat cheaper to apply, although with rising labour costs the gap between painting and galvanising is narrowing. However, the life of a zinc coating is generally considerably greater,

Fig. 63 – During erection of the Bosphorous Bridge in Turkey, the galvanised steel cables of the permanent structure were supplemented by galvanised wire mesh and other items for the construction walkway. (Courtesy of Zinc Development Association.)

and consequently maintenance costs are less. The overall cost of protection can be assessed only when both capital and maintenance costs, over the whole life demanded from the structure, are taken into consideration, and many factors are involved. These are discussed in Appendix E of BS 5493 and also in papers published by the Zinc Development Association [11], and in Metals and Materials [12].

Frequently it can be shown that when long life is required, a combination of galvanising or zinc spraying, with a paint system subsequently applied, gives the lowest overall cost. The recent suspension bridge over the Bosphorous is an example of this philosophy. The 16,000 tonnes of steel forming the structure of the bridge is protected by zinc spraying followed by a 4-coat paint system. The 5000 tonnes of cable was continuously galvanised, and the bridge parapets hot-dip galvanised after fabrication. The Forth Road Bridge was protected in a similar manner.

Fig. 64 – The Forth Road Bridge has galvanised steel cables, and the structure is zinc sprayed and painted. (Courtesy of Zinc Development Association.)

REFERENCES

[1] Frappa, A. L., Hot dip galvanising of constructural steels by the Technigalva or zinc/nickel procedure, Report on the general meeting of European General Galvanisers Association 1983, Zinc Development Association, London.

[2] Harper, S. and Browne, R. S., The range of alternative practices in galvanising steels containing silicon, *Proceedings of the 11th International Galvanising Conference 1976,* Zinc Development Association, London.

[3] Mathewson, C. H. (ed.), *Zinc: the metal, its alloys and compounds,* pp. 453–522, Reinhold, New York, 1959.

[4] The publications of the Hot Dip Galvanisers Association, The Zinc Development Association, London.

[5] The publications of the Zinc Institute, New York.

[6] *Design. Galvanising for Structural Steelwork 1979,* Galvanisers Association, London.

[7] Porter, F. C., *Metal Construction,* Oct., 1983, p. 606, and Nov., 1983, p. 676.

[8] Campbell, H. S., Sprayed aluminium and zinc protect against corrosion, Paper No. 6, *Association of Metal Sprayers Symposium, April 1982.*

[9] Britton, S. C. and Evans, W. R., *Journal of the Society of the Chemical Industry,* Vol. 51, 211T, 1932, and Vol. 55, 237T, 1936.

[10] Mayne, J. E. O. and Evans, W. R., Protection by paints richly pigmented with zinc dust, *Chemistry and Industry,* pp. 109–110, 1944.

[11] Porter, F. C., The economics of protection for steel products, *9th International Conference on Hot Dip Galvanising,* pp. 29–36, Zinc Development Association, London, 1970.

[12] Brace, A. W. and Porter, F. C., The economics of protection by zinc coatings, *Metals and Materials,* Vol. 2, No. 6, pp. 169–173, 1968.

13

Chemical applications of zinc

Metallic zinc and a number of zinc compounds are used for a variety of purposes in the chemical and metallurgical industries. The principal compounds are zinc oxide, zinc sulphate, zinc chloride and zinc phosphate.

13.1 ZINC DUST AND POWDER

Zinc dust, with its high metallic zinc content and large surface area, serves a number of applications in the chemical industry, mainly as a reducing, condensation and dehalogenating agent. One of its chief applications is in the production of zinc hydrosulphite, ZnS_2O_4, which is produced commercially from a mixture of zinc dust, sulphur dioxide and water:

$$Zn + 2SO_2 = ZnS_2O_4$$

The zinc hydrosulphite is frequently converted into the sodium salt by adding sodium hydroxide and removing the precipitated zinc hydroxide, which is frequently dried and used as a pigment. The hydrosulphite (either zinc or sodium) is used in textile and paper bleaching and also as a reducing agent in dyeing.

In the metallurgical field, zinc dust is used to precipitate gold from cyanide solutions, and to remove a number of metals during the purification stages in electrolytic zinc production.

In 1982 in the United Kingdom 10,000 tonnes of zinc dust were consumed in industrial applications, of which 3000 tonnes were used in paint formulations.

Zinc dust production

With most distillation methods the production of sometimes a considerable proportion of the zinc as a fine powder with a high metallic content is all too easy, and condensation systems are designed to reduce this proportion of blue powder to a minimum. Although the particles consist mainly of metallic zinc, they are invariably coated with a thin film of oxide, produced by the reversal of the reduction reaction

$$ZnO + CO = Zn + CO_2$$

This film makes it almost impossible to coalesce the zinc with even sophisticated melting methods, and consequently the blue powder has generally to be returned to the smelting operation.

A relatively small market for some of the powder has always existed, since, with its high metallic zinc content and fine state of division, it has found application in a number of industries. To satisfy this market, certain horizontal plants fitted steel canisters to some retorts instead of the orthodox condenser, so that the zinc vapour was rapidly cooled and condensed almost entirely as powder. In some vertical retort plants the vapours leaving the condenser were passed immediately into steel cyclones, in which the uncondensed zinc was cooled and caught as a saleable dust. At other plants, metallic zinc, sometimes in the form of dross or scrap, is distilled in oil-fired or electrically heated crucibles, and again the vapour shock-chilled to form dust.

One of the best methods of producing dust under closely controlled conditions was developed by the Imperial Smelting Corporation at Avonmouth. A vertical retort is used, square in section, 0.27 m² in area. The retort is filled with 20–30 mm coke to a height of 2.5 m and the electrical resistance of this coke is used to provide the heat necessary to distil molten zinc, which is fed into the top of the retort through orifices at the bottom of four crucibles, so placed as to give reasonably even distribution of metal across the retort. An inclined graphite rod enters the top of the retort and is buried in the coke, and graphite slabs on which the coke rests at the bottom provide the electrical connections. Under steady conditions the furnace consumes approximately 250 kW. The rates of feed and power applied are adjusted to distil approximately 7 tonnes of zinc per day, leaving 5–10 per cent of the feed unvaporised, so as to carry out unvolatilised impurities, such as zinc, iron and lead, through a run-off at the bottom of the retort. The vapour leaving the retort passes into a series of three steel tanks which are filled with nitrogen and provide the shock-chilling necessary to produce dust. This passes into a cyclone cooler, a high efficiency cyclone, a blower and then a sealed bag filter. The gas which is now cool and free from dust is fed back to nozzles surrounding the vapour entry, and provides the main shock-chilling effect. By varying the volume circulated between 5–10 m³/min, some control of the particle size of the dust can be made. Approximately 1000 kWh are required to produce 1 tonne of dust [1].

For some uses, such as in pyrotechnics and the Schori type of zinc spraying pistol, a relatively coarse dust is required, and this is generally made by impinging jets of high pressure air on a stream of molten zinc. In the high turbulence of the air stream, the zinc breaks up into fine droplets which chill and form irregularly shaped particles which are caught in a bag filter. The particles are always coarser than dust formed by distillation and have generally a mean particle size of approximately 70 μm.

A growing use of zinc powder is in the production of alkaline zinc–manganese dioxide primary cells. These cells cost more than the well known Leclanché cells with a zinc can, but are more economical for applications demanding a continuous high current discharge, e.g. for portable tape recorders and various toys with electric motors. The cells have a steel can with a thick layer of manganese dioxide on the inside, and inside that a cylindrical separator enclosing the powder anode in the form of a paste. A metal nail current collector forms the centre of the package and connects the anode with the steel can. The manganese dioxide cathode makes contact with the cell cap, which has a central projection. The cells are thus interchangeable with Leclanché cells of the same size, though their energy capacity is about double and they have a longer shelf-life.

Zinc powder is also used in miniature cells (button cells) with positive electrodes consisting of mercuric oxide for a stable voltage, or silver oxide to supply pulsed current. These cells are used in such applications as watches, calculators and hearing aids.

Another type of small cell using zinc powder has a porous cathode coated with a catalyst which allows air to act as the depolariser, i.e. to combine with the hydrogen that is formed in the current-generating reaction. Manganese dioxide is the depolariser in the Leclanché type cells. These zinc–air cells have a high energy capacity, high power and constant output.

13.2 ZINC OXIDE

Zinc oxide is commercially the most important chemical compound of zinc, and is used to a considerable extent in the rubber, ceramic, paint and other industries. In 1982, 19,000 tonnes of zinc oxide were used in the United Kingdom, the allocation being as in Table 26.

Table 26

Zinc oxide consumption in UK 1982

Rubber	50%
Paints	7%
Ceramics	7%
Chemical uses	36%

Zinc oxide has a number of remarkable physical properties which arise primarily from its structure, the lattice of which is hexagonal, of the wurtzite type. It can be considered to be built up of tetrahedra consisting of a zinc atom surrounded by four oxygen atoms and an oxygen atom surrounded by four zinc.

The bonding is not uniform, and although mainly ionic, covalent bonding occurs along the c axis, so that the tetrahedral structure composing the lattice is not entirely symmetrical [2]. Another feature of the lattice is that it is relatively open, since the zinc and oxygen atoms can be considered to occupy only 44 per cent of the volume, leaving relatively large spaces (0.095 nm radius) [3]. As a result other metal atoms can enter the lattice at elevated temperatures and introduce marked changes in properties.

By modifying the conditions of formation, or by subsequent heat treatment, commercial oxides can be produced in a variety of crystal forms, the main types being short needles, long acicular crystals, fourlings and flakes. Although many crystalline compounds and metals exhibit twinning consisting of two lattices growing from a common junction, zinc oxide is the only inorganic compound which readily forms fourlings, consisting of four acicular spines united at a common base. Flakes are formed by the development of plates continuous with the spines [4]. Since by controlling the conditions of formation these varieties of crystal structure can be produced at will, a wide variety of types of zinc oxide with differing physical properties are commercially available, including materials with a particle size varying from 0.1–5 μm. Micrographs of typical acicular and nodular commercial oxides are shown in Figs. 65 and 66. For rubber compounding nodular type oxide is generally used, whereas in the paint industry the

Fig. 65 – Zinc oxide, acicular. (Magnification $\times$ 10,000.)

Fig. 66 – Zinc oxide, nodular. (Magnification ×10,000.)

acicular type is preferred. Since zinc oxide displays a high degree of reactivity with both acids and alkalis, and since particles can be produced in a wide range of surface types and areas, it finds application in a number of chemical processes.

It is a useful dehydrogenation catalyst, and is used in the conversion of aliphatic hydrocarbons to unsaturated aromatics, and also in the synthesis of methanol from carbon monoxide and hydrogen. Another important application is in the curing of rubber; and in automobile tyres, its high heat conductivity and heat capacity reduce the tendency to develop excessive temperatures during flexing and distortion.

Although, as normally manufactured, zinc oxide is white, colour changes can be induced by introducing various ions into the lattice. Because of its whiteness and high refractive index (2.008) it was at one time widely used as a pigment in paints for both indoor and outdoor use, but it has now been largely displaced for this purpose by the titanium dioxide range of pigments. Since it strongly absorbs ultra-violet light it is a constituent of most suntan lotions and other cosmetics.

In addition to its more orthodox chemical and physical properties, which led to its use in the applications briefly outlined above, zinc oxide has a number of special properties which are beginning to be applied in more sophisticated fields. By suitable heat treatment, which multiplies the number of imperfections in the lattice, electrical conductivity can be increased and semiconductor properties developed. Heating for a short time in a reducing atmosphere at 900°C removes oxygen atoms, leaving excess zinc atoms which migrate to interstitial sites. These carry two positive charges, having released two electrons which cause the increase in conductivity. Imperfections can be produced in other ways, such as by the introduction of gallium oxide into the lattice, and again semiconductor properties are developed.

Zinc oxide also exhibits photoconductive properties, since under the action of light a marked increase in its electrical conductivity occurs due, it is believed, to the release of electrons from interstitial atoms. By heating in hydrogen to produce excess zinc in the lattice, a peak effect with white light at 420 nm can be produced, and this effect is utilised in photocopying. In this process, a paper coated with treated zinc oxide dispersed in a binder of high dielectric strength is charged to a potential of 500 volts which will be held as long as the paper is not illuminated. If an image is then projected on to the sheet, the charge on the illuminated areas will leak through the resin to the paper, but will be retained on the dark areas. The sheet is then immersed in a liquid containing a suitable dye stuff which migrates to the charged areas and is retained, the density being proportional to the charge. A positive replica of the original image is obtained, which can be fixed by heat treatment or other means, to give high quality replicas rapidly and cheaply. This Electrofax process has provided a major market for zinc oxide.

Zinc oxide which has been heated in a reducing atmosphere or to which have been added small quantities of other elements can act as a phosphor, converting ultra-violet light, X-rays or television cathode rays into light of various colours in the visual spectrum. Under certain conditions such phosphors also show electroluminescence and, in an alternating current field, can emit light.

Zinc oxide has a variety of remarkable properties and the study and application of these had led to advances in a number of fields of technology. A summary of its properties is given in the publication *Zinc Oxide: Properties and Applications* [5].

Zinc oxide production

Zinc occurs in many copper ores, and zinc oxide can be recovered form the furnace flues when such ores are smelted. It was therefore known relatively early in man's metallurgical history, and its medicinal and other properties recognised. 'Tutia' is one of the early names for zinc oxide, and an old Persian proverb states 'The dust of a flock of sheep is tutia to the eyes of a hungry wolf'.

There is evidence that in the second century BC, if not earlier, the oxide was made deliberately in small hearth furnaces in which zinc ores mixed with charcoal were heated, forced draught being provided by bellows. Zinc oxide was reduced and the metal volatilised to burn above the bed, forming a fume which was collected in a settling chamber attached to the furnace.

Today, zinc oxide is produced by two thermal methods and also chemically. The two thermal methods are known as the Direct and Indirect, depending upon whether the raw material was zinc oxide in the form of calcined ore or drosses, or metallic zinc.

As usual in the zinc industry a number of different types of furnace have been developed to exploit each of the thermal methods.

Direct processes

Much of the direct oxide produced today is made by heating a mixture of calcined or sintered blende and anthracite on fixed cast iron grates, the required proportions lying generally within the limits 60–70 per cent calcine, 40–30 per cent anthracite. The well-mixed charge is fed into a hot furnace and spread evenly over the grate to a depth which varies with the density of the charge, but is usually 400 mm. After the temperature of the top of the charge has reached that of the furnace arch, air is blown up through the bed and combustion of the top layers begins. Due to the excess of carbon in the bed, conditions are such that reduction of zinc oxide takes place and metallic zinc is vaporised. The gases leaving the bed, consisting of nitrogen, carbon monoxide and zinc vapour, pass to a combustion chamber where air is added to oxidise the zinc, and by controlling the conditions, oxide of varying particle size and shape to suit market demands can be produced. If fine particle size oxide is desired, air is admitted through a number of jets causing turbulence in a relatively small combustion chamber. The dust laden gas is then cooled as rapidly as possible, by the use of excess air or water sprays in the flue leading from the combustion chamber. If coarser material is required, the rate of cooling is reduced and the increased time at high temperature encourages particle growth. Acicularity is developed if a large combustion chamber is used and the zinc vapour burnt in excess air with little turbulence.

In the 1920s the New Jersey Zinc Company modified the process and used a travelling grate similar to that employed in chain grate stoker installations. First a layer of coal briquettes was fed on the grate and then the main layer of calcine–coal briquettes. The combustion of the lower layer of coal briquettes as the grate passed along the furnace provided most of the heat required to reduce the zinc oxide in the upper layer.

Rotary furnaces, as used by the Eagle Picher Company, are fed with a charge of calcined blende and anthracite, which is introduced at the firing end and passes down the kiln concurrently with the heating flame.

Yet another variation is used by the St. Joseph Lead Company, where the heat necessary for reduction and distillation is supplied electrically using the resistance of a charge of sintered blende and coke. The furnaces used are similar in principle to those producing zinc metal, but, for oxide production, the zinc vapour mixed with the accompanying carbon monoxide leaves the furnace through ports and is oxidised immediately in manifolds where controlled additions of air are made. By varying the rate of oxidation and subsequent cooling, a wide range of types of oxide particle can again be produced.

Indirect processes

The purest and finest grades of zinc oxide are made from a feed of metallic zinc which is boiled and the vapour oxidised, producing oxide which can thus be made free from the small quantities of impurities which originate in the calcine used as raw material in the direct process. Several different types of furnace are used to distil the feed metal. For many years gas- or oil-fired horizontal retorts were employed but these have been largely displaced by more efficient methods of distillation. Several companies use electric arc heating to vaporise the metal. The New Jersey Zinc Company produces oxide from furnaces employing the carborundum tray columns which it developed originally as reflux units to produce high grade metal. In yet another method, metal is fed into small rotary furnaces, in which it is melted and vaporised. The vapour is oxidized by air introduced above the bath, and the heat from the reaction is sufficient to distil further additions of metal as the furnace becomes thermally self-supporting. Whilst the process is attractive in that little heat has to be supplied, it has the disadvantage that it is difficult to adjust the conditions of oxidation and obtain rapid cooling of the oxide. Thus some of the control of particle size and shape, which the other methods of distillation permit, is lost.

Conditioning of zinc oxide

Before sale, zinc oxide is subjected to a number of conditioning steps. Screening and passage through a high-speed pulverising mill is almost invariably used. Direct oxide is frequently reheated in a kiln, to temperatures between $700°$ and $770°C$ in a near neutral atmosphere, to improve the colour and to reduce the sulphate content. Kiln treatment is also sometimes used to cause growth of the finer particles and thus reduce the size range. To improve the surface characteristics, some oxide is coated with films of stearic or other fatty acids to increase the rate of dispersion when compounded in paint or rubber.

The particle size of most commercial zinc oxide ranges from $0.01-1$ μm with the average generally about 0.25 μm. The specific surface area lies in the range $5-15$ m^2/g, but the range for indirect oxide is somewhat less.

Chemical process zinc oxide

When sodium hydrosulphite is made by treating zinc hydrosulphite with sodium hydroxide, zinc hydroxide is precipitated. After washing, it can be heated to form chemical process zinc oxide. The material lacks the crystalline form necessary in a paint pigment, but is suitable for incorporating in rubber, and as a material for conversion to zinc chromate, zinc phosphate, etc.

13.3 ZINC SULPHATE

It can be claimed that zinc sulphate is the most important zinc salt, since all electrolytic zinc is deposited from sulphate solutions, but in addition to this use, a considerable amount is used in general industry. In viscose rayon manufacture it is added to the precipitating bath to promote crenulation, and it is also used in textile dyeing and printing. Another important application is in agriculture and animal nutrition where it is used to correct zinc deficiency.

It is prepared mainly from zinc residuals, such as galvanising drosses and skimmings, by leaching with sulphuric acid followed by purification and crystallisation.

Zinc sulphate forms three common hydrates, $ZnSO_4.7H_2O$, $ZnSO_4.6H_2O$ and $ZnSO_4.H_2O$. It is generally sold in the monohydrate form, produced by evaporating high hydrates to dryness.

13.4 ZINC SULPHIDE

Although zinc sulphide occurs as the mineral sphalerite, which is the main source of zinc production, the mineral itself is always too impure to be used in industry as such, and when zinc sulphide is required it is always produced from other zinc salts — usually the sulphates — generally by passing hydrogen sulphide into a buffered solution of the salt.

After calcination, followed by quenching in water, zinc sulphide can be used as a white pigment, with a high refractive index (2.368 in sphalerite form) and fine particle size. In the past a mixture of zinc sulphide and barium sulphate, known as lithopone, found considerable use as a pigment. The mixture was formed by double decomposition from solutions of zinc sulphate and barium sulphide (produced by reducing barium sulphate with carbon):

$$ZnSO_4 + BaS = ZnS + BaSO_4$$

To obtain the necessary optical properties the mixed precipitate was heated to 700°C in kilns in the absence of air. Although zinc sulphide and lithopone were once important pigments they have now been largely replaced by titanium dioxide.

Zinc sulphide is an important phosphor, and can be made to phosphoresce and fluoresce in many different colours by adding small controlled quantities of activators such as copper, manganese or silver to pure zinc sulphide and heating the mixture to a high temperature (800–1200°C). The most common zinc sulphide phosphor uses a copper activator and gives a bright green luminescence.

13.5 ZINC CHLORIDE

Zinc chloride, another important commercial salt, is mainly produced by dissolving zinc residuals in hydrochloric acid, purifying the solution and then evaporating until the zinc chloride content rises to 47.4 per cent. Pure anhydrous zinc chloride can be prepared by treating molten zinc with dry hydrochloric acid at 700°C or by reacting dry zinc sulphide with chlorine.

Zinc chloride is very soluble in water and highly deliquescent. It is also soluble in a number of organic solvents such as alcohols and ethers. The fused salt melts readily to a clear liquid and can be distilled readily at 900°C. Solutions hydrolyse to form oxychlorides and cannot be evaporated to dryness without considerable decomposition, with the formation of hydrochloric acid. Concentrated solutions are acidic and dissolve many organic materials such as starch and cellulose.

Zinc chloride forms double salts with many other metallic halides. The double salts with ammonium chloride, $ZnCl_2 2NH_4Cl$ and $ZnCl_2 3NH_4Cl$, are readily formed and are used widely, mainly as fluxing agents. Fused zinc chloride itself dissolves many metallic oxides and is the basis of a number of fluxes used in metallurgy. It is also used in aqueous solution as a wood preservative either alone or with sodium chromate or phenol. It is employed as a mordant in printing and dyeing textiles, and also in mercerising cotton. Other uses are in oil refining, galvanising and in the manufacture of dry batteries.

13.6 OTHER ZINC SALTS

Zinc chromate
Several important corrosion-inhibiting pigments are compounds which contain zinc chromate in various proportions. Their effectiveness depends upon the slow release of chromate ions in contact with water, making them valuable as pigments for primers to be used on both ferrous and non-ferrous metals. However, the chromate ion is toxic and so these pigments are avoided where they might constitute a health hazard. Paints containing zinc dust or zinc phosphate may then be used instead.

Zinc phosphate
The system $ZnO–P_2O_5–H_2O$ is complex and a number of compounds can be produced. All contain some water of hydration, but upon heating to 150°C all revert to the anhydrous salt $Zn_3(PO_4)_2$, which is relatively insoluble in water and in alcohol.

Zinc phosphate solutions are widely used to passivate steel and other metal surfaces and thus retard corrosion. The solution is usually prepared on site by adding phosphoric acid to a slurry of zinc oxide. The metal surface to be treated is cleaned and then immersed in or coated with the phosphate solution. On steel an adherent film composed of iron and zinc phosphates is formed, which is not only corrosion-resistant itself but also forms a good base for paint and other coatings. It has good lubricating properties and is used as a pretreatment for parts which must be heavily drawn or formed. Zinc phosphate coatings are also effective in improving paint adherence on aluminium and other metals.

Zinc phosphate has become important as an anticorrosive pigment for metal primers. It is non-toxic and has replaced red lead and zinc chromate in some applications.

Zinc fluoride

ZnF_2. Zinc fluoride is formed by adding hydrofluoric acid to a slurry of zinc oxide. The water of hydration can be removed by heating and the tendency to form basic fluorides is considerably less than with the other zinc halides. The fluoride is relatively insoluble in water and forms no acid fluorides. It is thermally stable. It is used to a small extent in making phosphors, in galvanising fluxes, as a wood preservative and in ceramic products.

Zinc phosphide

In damp air, zinc phosphide, Zn_3P_2, slowly releases phosphine and is used as a poison for rats and other rodents. It is produced by heating zinc and phosphorus together in an electric furnace at $700^\circ C$. The product can be purified by sublimation. It is insoluble in water and alcohol.

Zinc borates

Zinc borates, $3ZnO.2B_2O_3$, of varying composition are made by heating the oxides together with an activator such as manganese. They are used for their fluorescent and thermoluminescent properties. In solution, zinc borates are made by mixing a solution of borax and zinc sulphate, or reacting zinc oxide or carbonate with boric acid. Such solutions are used for fireproofing textiles, and as fungi and mildew inhibitors.

Zinc soaps

A range of zinc soaps can be made, with the general formula $(RCOO)_2Zn$, R being an organic radical, usually aliphatic in type, containing at least six carbon atoms. The soaps are made either by precipitation, by mixing solutions of zinc sulphate and a soluble sodium or potassium soap, or by fusing zinc oxide with the appropriate organic acid. Zinc soaps are insoluble in water, but dissolve in nonpolar liquids such as benzene. The most widely used zinc soap is the stearate.

Zinc soaps are used in the paint industry, their main function being to hasten the drying of oil, varnish or enamel films; but to do this they must be combined with other driers such as those containing cobalt or manganese.

The zinc salts of the saturated acids such as stearic, $C_{17}H_{35}.COOH$, are widely used in the plastics industry as plasticisers, internal lubricants and stabilizers. In the rubber industry they act as softeners and vulcanising activators and they also shorten the curing time. The zinc soaps are also used as fungicides and as wood preservatives — particularly zinc napthenate. Zinc stearate is used as a lubricant in powder metallurgy and in drawing and forming operations.

Thio-salts

Zinc salts of various thio-acids such as zinc dibutyldithiocarbamate, form a small but important group which serves a number of industries. They act as accelerators for rubber vulcanisation, and are also used as fungicides for crop protection, and in the paper industry for controlling sliming. Zinc dialkyldithiophosphates are added to engine oils as antioxidants and as detergents.

REFERENCES

[1] Newton, D. S., Electrothermal process for zinc dust production, *World Symposium on Mining and Metallurgy of Lead and Zinc,* Vol. 2, p. 995, American Institute of Mining and Metallurgical Engineers, 1970.

[2] Johnson, *Physics Review,* Vol. 57, p. 613, 1940.

[3] Bragg, *Transactions of Faraday Society,* March 25, 1929.

[4] Cowley Rees and Spink, *Proceedings of Physics Society,* Vol. 64A, p. 609, 1951, and Vol. 64B, p. 638, 1951.

[5] Brown, H. E., *Zinc oxide: Properties and Applications,* International Lead Zinc Research Organization, New York, 1976.

14

Biological significance of zinc

The presence of small quantities of zinc is essential to both animal and plant life. 'If one had to choose the most important (micronutrient) it would be zinc' [1]. The human body contains on average approximately 2 g of zinc, present in most organs at concentrations lying between 20 and 30 μg of ash. It has been estimated that most normal diets contain 10–15 mg of zinc per day, most of which originates in the protein intake, and is excreted in the faeces. Its significance in pathological processes is not yet clear, but it has been shown that zinc is an integral part of the carbonic anhydrase molecule which is an essential part for the exchange of carbon dioxide [2]. It is also present in at least six other enzymes in the human body. The utilisation of nitrogen and sulphur in the animal body requires zinc. It also plays a critical role in the processes of cell division and growth, and is also involved in the metabolism of the pituitary and adrenal glands. It is also essential for spermatogenesis, formation of ova, and in the nutrition of the foetus [3].

Claims have been made that in certain countries, particularly the Middle East, nutritional dwarfism occurs associated with zinc deficiency. The salient features were growth retardation, absent sexual development, iron deficiency, anaemia, and geophagia. Most cases occurred in areas where the diet consisted mainly or exclusively of wholemeal bread, containing a high level of phytate which tends to make zinc inactive. It was demonstrated that the conditions could be ameliorated by regular additions of zinc sulphate [4].

Whilst the ingestion of excess quantities of zinc in food can create discomfort, due to malaise, colic and diarrhoea, the effect is transitory, and few fatal cases have been reported. The Environmental Protection Agency in the USA, the World Health Organisation and also the Australian and Canadian public health agencies recommend a limit to zinc concentration in domestic water of 5 mg/litre.

Zinc salts have little effect on the human skin (with the exception of zinc chromate, probably because of the chromate radical); in fact zinc oxide has been used as an ointment since before the Christian era.

Under certain circumstances, the inhalation of fresh zinc oxide fumes arising from distillation processes or brass melting carried out with inadequate ventilation causes acute symptoms of malaise, with an increase in body temperature, a cough which may cause vomiting, excessive salivation and severe headaches. These symptoms are known as 'zinc shakes', 'brass chills' or 'metal fume fever'. Although they can also arise from the inhalation of magnesium oxide or cadmium oxide fumes, and in the case of cadmium can be fatal, permanent effects due to zinc or magnesium oxides are almost unknown, and recovery is rapid.

Any risk of such trouble can be avoided by ventilation which reduces the average concentration of zinc oxide in the atmosphere to 5 mg/m^3, or by the use of respirators if exposure to greater concentrations for short periods cannot be avoided. The effect of zinc oxide is not cumulative and does not appear to cause other respiratory disorders. Little appears to be known of the mechanism by which zinc oxide causes these symptoms, but it appears that only freshly formed zinc oxide can cause the symptoms specified.

Zinc in agriculture

In agriculture, the presence of trace quantities of available zinc (together with other elements such as iron, copper, manganese, boron and molybdenum) is essential for healthy growth and good cropping. It is effective in such small quantities because it forms part of the constitution of the enzymes which regulate plant life. Thus, zinc is needed for the production of auxins, the growth-promoting substances which control the growth of shoots.

Most soils contain sufficient available zinc to supply plant needs, but there are areas in the world where even the small quantity of zinc required is not naturally available. In such cases yields are low, seeds do not form and the crop may be a total failure. Many crops are affected in such conditions, but citrus fruits seem to be the most sensitive. Such plant diseases as 'mottle leaf' of citrus, 'white bud' of corn, 'little leaf' of apples or pears are readily recognisable evidence of zinc deficiency.

Although few cases of zinc deficiency in agriculture have been reported in the United Kingdom, wide areas are affected in the United States, India, Australia and in a number of other countries. In many areas the problem can be overcome by applications of zinc either in solid or in solution. The zinc can be added in a number of forms, but zinc sulphate, zinc oxide or the carbonate added in slurry form are the compounds most widely used [5].

REFERENCES

[1] Schroeder, H. A., *The Poisons Around Us. Toxic Metals in Food, Air and Water,* Indiana University Press, Bloomington, Ind., USA, 1974.

[2] Keilin, D. and Mann, T., *Biochemical Journal,* Vol. 34, pp. 1163–1176, 1940.

[3] Sandstead, H. H., Some trace elements essential for human nutrition, *Prog. Food Nutr. Sc.,* Vol. 1(6), p. 371, 1975.

[4] Halstead, Ronaghy, Abadi, *et al., American Journal of Medicine,* Vol. 53, p. 277, 1972.

[5] *Zinc in Crop Nutrition,* Zinc Development Association, London, 1971. Also *Zinc in Animal Nutrition,* American Zinc Institute, New York, 1964.

Appendices

1. ENERGY BALANCE OF ELECTROLYTIC AND BLAST FURNACE PROCESSES

A comparison of the energy required to produce zinc by typical operation of the blast furnace and electrolytic routes was given by Hopkin and Richards [1] and Hopkin [2] as follows:

Table A.1

Specific energy consumption of the zinc–lead blast furnace

Products: 1 tonne Zn4; 0.5 tonnes Pb; 2 tonnes H_2SO_4

Unit process	Specific consumption			
	Electricity (kWh)	Coke (kg)	Oil (kg)	GJ[a]
Concentrate handling	12	–	–	0.04
Sinter roast	145	–	20	1.32
Furnace	368	1030	36	31.66
Cadmium	–	–	1.4	0.06
Acid	150	–	–	0.54
Totals	675	1030	57.4	
MJ/unit (net)	3.6	28.0	40	
GJ equivalent	2.43	28.9	2.30	33.6
High grade concentrates				
1.76 t zinc concentrate				7.7
0.62 t lead concentrate				1.2
Secondary energy subtotal				42.5
Power station thermal losses @ 67%				4.9
Coke oven losses @ 15.5%				5.2
Credit for unused furnace gas (LCV)				(4)
Total primary energy requirement				48.6
Credit for 0.5 t lead (blast furnace route)				(8)
Total primary energy requirement, 1 t zinc				40.6

[a]Net calorific values used.

Zinc—lead blast furnace

Typical results were taken from operation of a 17.2 m² hearth area furnace treating zinc concentrates (analysis, assumed, zinc 56 per cent, lead 2.5 per cent) and lead concentrates (lead 60 per cent, zinc 10 per cent), burning 210 tonnes of coke per day and producing daily 256 tonnes of zinc and 128 tonnes of lead.

Table A.1 details the specific energy consumption to produce 1 tonne of Grade 4 zinc, and Table A.2 gives the additional energy requirement to produce Grade 1 zinc. The majority of furnace operators upgrade only a proportion of Grade 4 metal.

Table A.2

Specific energy consumption of zinc refining processes

Refluxing process to 1 tonne Grade 1 zinc Lead and cadmium columns		GJ
Electricity	22 kWh	0.08
Oil	160 kg	6.4
Secondary energy subtotal		6.5
Generation losses @ 67%		0.16
Primary energy consumption		6.6

Electrolytic process

For the purpose of comparison, the operation of a typical electrolytic plant was assumed as typified in Reference 3, and summarised in Table A.3.

Comparison of energy requirements of the two processes

At first sight it would appear from the foregoing tables that the primary energy requirements of the two processes are similar. It should be remembered however, that in the case quoted, for every tonne of zinc, the blast furnace produces also 0.5 tonne of lead. Thus, the blast furnace is entitled to a credit from the production of this lead of 8 GJ. When this is deducted, the saving in net energy in the blast furnace case, over electrolytic treatment, is considerable.

REFERENCES

[1] Hopkin, W. and Richards, A. W., *J. Metals,* Vol. 30(11), pp. 12–17, 1978.
[2] Hopkin, W., *Symposium on Energy Considerations in Electrolytic Processes, SCI and IMM, Newcastle upon Tyne,* pp. 43–59, July 1980.
[3] *AIME World Symposium on Metallurgy & Environmental Control of Pb, Zn and Sn, AIME, Las Vegas,* p. 28, 1980.

Table A.3

Specific energy consumption of the electrolytic zinc process
(General Practice Ref. [3], p. 28)

Products 1 tonne Zn1, 1.62 t H_2SO_4 (net)

Unit process	Electricity (kWh)	Oil/gas (GJ equivalent)	GJ
Concentrate drying Roasting	120		0.43
Leaching Solution purification Residue treatment	190		0.68
Electrolysis	3580		12.9
Melting, casting, zinc dust	134		0.48
Steam plant		$1.35\pm^a$	1.35
Acid	240		0.86
Miscellaneous	70		0.25
Totals	4334	1.35	
MJ/unit (net)	3.6		
GJ equivalent	15.6	1.35	16.95
High grade concentrate, 2.0 t			8.7
Secondary energy subtotal			25.65
Power station thermal losses (67%)			31.7
Total primary energy requirement			57.3

^{a}Net calorific value.

2. ANALYTICAL CONTROL IN THE ZINC INDUSTRY

The development of the continuous methods of zinc production now in use, both electrolytic and pyrometallurgical, has made many new demands upon the analytical services. The successful solution of the problems thus posed has contributed significantly to the growth and increased efficiency of the industry.

The classical chemical and simple instrumental (e.g. colorimetric) methods which served adequately in the days of the older batch processes, proved too slow and laborious for modern smelter control. The introduction of sophisticated instrumentation, using physical or physical–chemical principles, provided techniques able to produce large numbers of results, on often complex materials, quickly, and yet with an acceptable degree of accuracy.

The modern smelter laboratory will thus be equipped not only to perform a variety of chemical analyses using classical methods, but will also employ a selection of instruments, ranging from the ubiquitous pH meter, and simple spectrophotometers, to atomic absorption spectrophotometers, photographic and direct reading emission spectrographs, X-ray fluorescent spectrometers, and possibly a plasma emission spectrometer.

The polarograph, which was widely used for a number of years, has now been largely superseded by the atomic absorption spectrophotometer. The latter, in its various forms, can cope with virtually all analyses for which the polarograph was previously employed.

Classical methods are still used for the analysis of concentrates and raw materials where payment is involved. Even here the time-honoured ferrocyanide method for determining zinc content is giving way to ones using E.D.T.A. as titrant. Atomic absorption methods are of value for determining the content of such values as cadmium, and are tending to supplant the fire assay procedure for silver and gold.

Chemical methods are also widely used in the preparation of standards for the calibration of the more sophisticated instrumental methods, which can then be used for routine analysis. A number of standards, the composition of which has been determined and agreed by several established laboratories, are now commercially available, and can relieve the laborious work of calibration.

In order to control a modern blast furnace and its associated plants producing zinc and lead, analysis of the input (sinter) and residue slag is required at least every two hours and sometimes more frequently. This is also the case with an electrolytic plant, but in this case almost continuous monitoring of the solutions at various stages is also essential. For those purposes, X-ray fluorescence spectrometry, together with a dedicated computer for control and evaluation, is widely used. On the other hand the metal products are more suited to analyses by emission spectrography or spectrometry. For the most rapid results an automated spectrometer is desirable.

The more automatic analytical methods are not always suitable for use with materials such as hygiene samples, effluents, and drosses. The analysis of these materials is usually carried out by atomic absorption techniques, or with the plasma emission spectrometer, supplemented by traditional methods of analysis.

Many of the analytical problems connected with the smelter are also of relevance to the mine and concentrator, and similar instrumentation and techniques are being employed in the laboratories of such plants. The atomic absorption spectrometer has found especially wide application, and is particularly valuable in prospecting work, where it can be used in the field, as can radio-isotope X-ray fluorescence methods. The latter finds use in on-line monitoring of flotation processes.

3. THE IMPORTANCE OF HYDROGEN OVERVOLTAGE IN ZINC ELECTROLYSIS, G. C. Bratt

The overall cell reaction is frequently written as:

$$ZnSO_4 + H_2O = Zn + H_2SO_4 + O \tag{i}$$

but cursory inspection of this equation suggests that some of the products may react further as follows:

$$Zn + H_2SO_4 = ZnSO_4 + H_2 \tag{ii}$$

and thus partially nullify the desired effect of the reaction in equation (i).

At the normal cathode potentials employed (0.8 V), both zinc and hydrogen would be codeposited at the cathode, but at differing rates. It is convenient to consider hydrogen which may have arisen from reaction (ii) to have been co-deposited with the zinc. Hence in examining the efficiency of the process the two partial equations:

$$Zn^{2+} + 2e = Zn \tag{iii}$$

$$H^+ + e = H \tag{iv}$$

can be considered as occurring simultaneously at the same surface.

The rates of these electrochemical processes may be expressed in terms of the current density, i, by use of the Tafel relationship as follows:

$$i^{Zn} = i_0{}^{Zn} \exp\left(-\alpha^{Zn}\eta^{Zn} F/RT\right)$$

$$i^H = i_0^H \exp\left(-\alpha^H \eta^H F/RT\right)$$

where

i is the actual current density

i_0 is the exchange current density (the magnitude of the equal and opposite current densities flowing at equilibrium conditions) and is determined by the composition of the electrolyte

α is the exchange coefficient

T is the temperature

F, R are constants

η is the overvoltage which is a measure of the additional effort required to operate the process at any current density, i, greater than i_0

The superscripts Zn, H, refer to the ions involved.

The total current density $I = i^{Zn} + i^H$, and the current efficiency, CE, for zinc deposition is:

$$CE = \frac{100 i^{Zn}}{I} = \frac{100(I - i^H)}{I}$$

For industrial conditions of approximately constant cathode potential (for which η^H becomes constant) the equation above may be rearranged to give

$$CE = \frac{100\,\{I - i_0^H \exp\,(-\alpha^H \times \text{const.})\}}{I}$$

and since α^H is also almost constant, the current efficiency of the process is determined by i_0^H. This parameter, which defines the hydrogen overvoltage, is determined by the cathode composition and, since metal impurities may be deposited at the cathode from the electrolyte, by the electrolyte composition. Thus, a high degree of purification of the electrolyte is essential.

REFERENCE
Bratt, G. C., Impurity effects in the electrowinning of zinc and cadmium, *Electrochemical Technology*, Vol. 2, pp. 323–326, 1964.

4. SI UNITS IN THE INDUSTRIAL METALS SERIES

Quantity	SI unit	Recommended multiples	Other related units (or names)
Length	m (metre)	km m mm nm	$\text{Å} = 10^{-10}$ m
Area	m^2	mm^2	hectare (ha) $= 10^4$ m^2
Volume	m^3	mm^3	litre (1) $= 10^{-3}$ m^3
Time	s (second)	s ms ns	minute (min) hour (h) day (d)
Mass	kg	g mg μg	tonne (t) $= 10^3$ kg metric carat $= 2 \times 10^{-4}$ kg
Density	$\dfrac{kg}{m^3}$	$\dfrac{g}{cm^3}$	
Force	N (newton)	MN kN	dyne $= 10^{-5}$ N

Quantity	SI unit	Recommended multiples	Other related units (or names)
Impact strength	$\dfrac{J}{m^2}$	$\dfrac{kJ}{m^2}$	$\dfrac{J}{cm^2}$
Electric current	A (ampere)	mA	
Electric potential	V (volt)	MV kV mV	
Current density	$\dfrac{A}{m^2}$	$\dfrac{A}{cm^2}$ $\dfrac{A}{mm^2}$	
Resistance	Ω (ohm)	$m\Omega$ $\mu\Omega$	
Resistivity	Ωm	$\mu\Omega m$	$\mu\Omega cm = 10^{-8}\ \Omega m$

5. COMMON CONVERSION FACTORS

1 yard	= 0.9144 m
1 foot	= 0.3048 m
1 inch	= 25.4 mm
1 mil	= 0.001 in = 25.4 μm
1 yd^2	= 0.836 127 m^2
1 ft^2	= 0.092 903 m^2
1 in^2	= 6.4516 cm^2
1 yd^3	= 0.764 555 m^3
1 ft^3	= 28.3168 dm^3
1 in^3	= 16.3871 cm^3
1 gal (Imp.)	= 4.5461 litre or = 4.54609 dm^3
1 in^4	= 41.6231 cm^4 (Moment of Section)
1 troy oz.	= 31.1035 g
1 dwt (pennyweight)	= 1.555 17 g
1 oz (av)	= 28.3495 g

Quantity	SI unit	Recommended multiples	Other related units (or names)
1 lb	= 0.453 592 kg		
1 cwt	= 50.802 kg		
1 ton	= 1016.05 kg		
1 tonne (metric)	= 1000 kg		
1 lb/in^3	= 27.6799 g/cm^3		
1 lbf	= 4.448 22 N		
1 tonf	= 9964.02 N		
1 tonf/in^2	= 15.4443 MN/m^2 or = 15.4443 N/mm^2		
1 assay ton	= 32.6667 g		

Index